Kinetics of 5-Nucleotidase in serum of liver Cirrhotic and normal Iraqi individuals

Prof.Dr. Sami A. AL-Mudhaffar

Vivian S. Aboudi

CHAPTER ONE
INTRODUCTION

I. Enzymes and Diseases

A - General Considerations

Enzymes :

Enzymes are organic catalysts produced by living organism[1], differ from other catalysts in many features, such as, the chemical nature, mechanism of action, and the reaction kinetics.

Diseases :-

A biochemical derangement in the cell, was called a molecular disease, and applied first to sickle cell anemia[2], this disease can be traced to the Hb concentration[3]. Later, in recent years many other compounds have been related the other diseases, e.g., glucose to diabetes mellitus, uric acid to gout, bilirubin to jaundice[4]. since these biochemical constituents are synthesized and degraded in the cells, through the action of specific enzymes, one could relate these enzymes directly to various diseases rather than relating the disease indirectly to various biochemical constituents.

B - Historical

_ The enzymes role in clinical medicine is new in its systematic approach, expanding very rapidly in many respects, level determinations, kinetic measurements, automatic evaluation ...... etc.

1

Enzymes have been used in clinical medicine very long, before, purification, chemical characterization, and kinetic studies. The crude ferments used in medicine can be traced back to the 18th Century, gastric juice was used in treatment of ulcer diseases[5]. John Hunter was said to have used pancreatic juice in infected wounds[6]. Nelson and Mayer[7] put the enzyme pepsin to use in treatment of dyspepsia, papain was used effectively in treatment of tuberculous ulcers[7]. The first clinical trials of purified TR were for the treatment of cancer[8 - 11]. The first application of enzymes as diagnostic tool was made by H.D. Kay[12] in 1930, using serum AP for bone diseases, and obstructive jaundice diagnosis. Amylase level in serum have been used also in the diagnosis of acute pancreatitis reported by R. Elman[12], later the discovery of transaminases in 1937 by Gutman and Gutman[12], has added to the list of clinically important enzymes. Otto Warburg[12] in 1943 initiated the clinical role and the diagnostic tool of glycolytic enzymes. From 1943 and onwards their has been a tremendous interest in application of enzymes in medicine.

C - Rate of Enzymes Release from Cells

There is a variation in the activities of certain enzymes from one tissue to another[13], the rates at which enzymes leave their cells of origin will depend on the integrity of the cell, its metabolic activity, the molecular configuration, charge and binding of enzymes to mitochondria.

Zierler[14, 15], has found a constant and appreciable rate of release of aldolase from muscle into the surrounding medium, he stated also that, this rate is effected by many factors, e.g. pH, glucose level, and others.

D - Origin of Serum Enzymes

Several studies have been made, concerning the origin of serum enzymes from tissues or cells, the first of such studies was the exciting investigations of Schlamowitz[16] utilizing immunological techniques, to purify AP from bone and intestinal mucosa.

Vessell and Bearn[17], have studied the problem of the Origin of serum enzymes by electrophoretic methods.

The complete specificity of enzyme localization in a single tissue helps in the specific diagnosis for the abnormal activity of serum enzymes.

4

E - Fate of Serum Enzymes

Studies have been directed towards the determinations of the fate of different glycolytic enzymes after intravenous injection and means of affecting the rate of inactivation, disappearance, and removal from blood[18]. To study the role of liver and other organs in the inactivation of LDH and ICDH in serum, experiments were run on a series of dogs, which were splenectomized, nephrectomized, and hepatectomized, the results of these experiments have suggested that the presence of kidney, spleen, or liver are not necessary for a normal rate of removal of these two enzymes. As a result, the abnormal activity of ICDH in patients with hepatitis is not due to inefficient inactivation of this enzyme by damaged liver cells[19].

II. Distribution of Enzymes in Body Fluids

Enzymes are distributed in a majority of body fluids, the level of the enzymes in these fluids varies according to, the type of enzyme, its source, and the role of these fluids.

A - Blood Plasma

Blood normally contains only very small amounts of many enzymes which are of tissue origin, because of their proteinic nature, enzymes don't diffuse out of their cell, but in diseases enzymes level increase in blood plasma due to the change in the concentration of the enzyme and also due to the effect of the transport mechanism.

Many enzymes present in blood plasma as, amylase, lipase, prothrombin, AP, GOT, GPT,....etc., these enzymes are either plasma specific or non-specific[20].

Plasma - Specific Enzymes

Include those enzymes which were originated from certain organ and secreted in an active form into the plasma. Examples are those involved in the control of blood coagulation " prothrombin, plasminogen[20]" .... etc.

Plasma Non - Specific Enzymes

1.   Secreted enzymes, are found in plasma and are not necessary for the function of the plasma, such as, pancreatic, parotid alpha amylase[20]..... etc.

6

2.   Organ — specific Enzymes, these enzymes are asso-
ciated with a particular organ, so their level in blood
plasma, reflect the state of that organ e.g., SDH, G6Pase,
and AP of the bone[20].

3.   Cellular enzymes, include enzymes participated and
synthesized in different cells, such as, GOT, GPT, LDH,
MDH, which are ineffective in plasma[20].

B - Urine

Many enzymes are found in urine, including amylase,
pepsin, LDH, AP, TR, and lipase. The possible sources
of the enzymes in urine are, serum, kidney tissues, ery-
throcytes, microorganisms and leucocytes, but kidney
tissue is assumed to be the main source of the enzymes
which appear in urine[21].

The assay used for determination of enzymes in urine
is complicated by a number of factors, pH, inhibitors, and
others.

Recent evidence[22] indicates that urine's LDH levels
are valuable for differentiating benign from malignant
lesions in the urinary tract. LDH is also elevated in
the presence of renal and prostatic adenocarcinomas.

AcP activity is increased in in urine in the presence of damaged renal tissues. Transaminases activity can normally be detected in urine, as a rule GOT activity usually exceeds that of GPT[23].

C - Other Body Fluids

AcP activity has been studied in gastric fluid. The low pH was found to exert a profound inactivating effect. There was an elevation in AcP activity in patients with gastric carcinoma[24].

LDH is found in gastric juice aspirated from normal individuals, but only if the material is neutralized, since the enzyme is inactivated at a low pH[25]. Lipase activity has been detected in various gastro intestinal fluids as gastric juice, jejunal secretion, pancreatic secretion[26].

Transaminases, as GOT has activity in CSF, obtained from adults, on the other hand no transaminase activity is usually found in CSF of healthy infants and children[27].

Saliva contains high activity of LDH in subjects with dental carrier or other inflammatory processes in the oral cavity[28]. Saliva contains an amylase (ptyalin) which is secreted at approximately its optimum pH 6.9 with chloride ions which activate it[29]

III. Enzymes Applications in Medicine

Various enzymes are present in serum in the normal or abnormal condition. The level of the specified enzymes helps in the diagnosis of these conditions, high or low level is related to certain pathological situation. Accordingly it was decided to follow the level measurements of certain clinically important enzymes to help in the diagnosis and prognosis.

A - Clinically Important Enzyme's Level in Various Diseases

1. Liver Diseases: -

Human liver is rich in many of the clinically important enzymes, such as, LDH, GOT, GPT, AP. Activity determination of these enzymes is suitable for the diagnostic purposes and for following the progress of the disease .

Serum Transaminases

The transaminases are a group of enzymes that catalyze the transfer of an amino group from an alpha amino acid to an alpha keto acid. Normally almost all of the transaminases are localized within the cell and relatively small amounts of them circulate in serum.

The relative tissue distribution of GOT and GPT have been studied in man. Hepatic tissue possess the highest content of GPT, and heart contains the highest content of GOT[23].

If hepatic cells which are rich in transaminases are damaged or destroyed, these enzymes will be released in the circulation, and their activity in serum under these conditions is dependent on the relative rates at which the enzymes enter and leave the circulation. Serum transaminases activity increase in most hepatic dysfunction[30,31] in acute hepatitis, the activity of GOT is increased in serum 10 - 150 fold and GPT is increased 20 - 200 fold. In chronic hepatitis and cirrhosis of the liver, increased activities of transaminases are found in serum. Toxic liver damage, during the acute toxic injury produced by carbon tetrachloride, the largest elevations of transaminases activities are observed in serum, the activity increases to several thousands units. In obstructive jaundice, the enzyme pattern in serum varies with the progress of the condition, the transaminases increases up to 50 fold, in the first days, but the level is decreased within a few days, to the normal range.

Serum AP

AP is a term applied to a group of enzymes that catalyze the hydrolysis of organic phosphate esters at an alkaline pH, with the liberation of inorganic phosphate. The main source of AP in the normal adult are; skeleton, hepatobiliary system, and the intestinal tract[32]. Elevation of serum AP activity is observed frequently in various hepatobiliary diseases, similar elevations are observed in disorders of bone.

Serum LAP

This is a protolytic enzyme, which hydrolyze amino acids from the N – terminal and of a number of protein and polypeptides this enzymes is called LAP because it reacts most rapidly with leucine compounds, LAP has been demonstrated in all human tissues with high activity in the liver where it is localized in biliary epithlium[33], LAP is increased with cholestasis, hepatic infilteration or metastasis, the results are parallel to AP, it is frequently increased in those with juandice due to extra-hepatic biliary obstruction, caused by pancreatic carcinoma or due to hepatic metastasis[34].

Serum γ - GT

This enzyme catalyzes the transfer of the γ -
glutamyl group from γ - glutamyl peptides to other
peptides and L - amino acids, there is a high enzyme
activity in the kidney, but lower activity is found in
the liver and pancrease[35]. Abnormal high values of γ -GT
appear in diseases of liver, biliary tract and pancrease[36,37].
In acute hepatitis γ - GT rarely reaches 10 - fold the
normal level, in chronic hepatitis and cirrhosis, γ -GT
has higher value than GOT.

2. Heart Diseases

Assays of certain enzymes in serum aid in the
diagnosis of heart diseases, in particular, the early
diagnosis of myocardial infarction, add to the physical
and electrocardiographic studies.

Myocardial Infarction :-

Kerman and Wroblewsky[38] reported that patients
with recent myocardial infarction had elevated activity
of GOT in their sera, due to high concentration of GOT
contained in heart muscle. The degree and duration of
the serum GOT elevation appears dependent upon the
magnitude of the infarction. LDH is elevated in myocardial infarction within about 72 hourse after infarction and may remain high for over a week[39], the magnitude

of the increase may be valuable for detecting minor degrees of myocardial injury that would tend to elevate transaminases levels remarkebly. The determination of CPK activity has also a practical value in the diagnosis of myocardial infarction, the rise in CPK activity in serum can be used for the early diagnosis of myocardial infarction.

3. Renal Diseases

Enzyme activity measurement is important in the clinical diagnosis of renal diseases. Although opinions differ recent evidence indicates that urine LDH levels are valuable for differentiating benign from malignant lesion in the urinary tract, such high values are noted, in patients with inflamation of the urinary tract. Patients with carbinoma of the bladder, chronic pyolonephritis, lupus nephritis, exhibit high out put of LDH[28]. In case of acute kidney impairments, there is high activity of AcP in serum and urine[24].

4. Tumour Growth

The influence of tumour growth on patterns of enzyme activity has been studied for many years. The possibility that the growth of tumour in one tissue organ may alter the enzyme pattern in distant tissues, particularly of tissue that are concerned with the elaboration of enzymes into the circulation. Brahn[40] first reported the decrease of liver catalase in 1916 in patients with various types of cancer. An interesting example of the effect of tumour growth in man on enzyme activity at a distant is the increase rate of glycolysis in patients with cancer[41].

5. Mental Diseases

Specific enzymic defect occurs in neuropathological conditions, but very little evidence is available to indicate the exact nature of such metabolic lesions, e.g., in myasthenia graris there appears to be an insufficiency of acetyl choline at the motor end plate, due to inadequate synthesis[42] or too rapid destruction by excess AChE.

6. Diseases of the Eye

Little information has been known regarding the specific enzymatic processes in eye, because of the specialzied nature and the relatively small size of the eye.

Normal tears contains mucolytic enzyme lysozyme, this lysozyme is typically lacking in the disease keratoconjunctivitis sicca or " dry eye " occuring predominantly in postmenopausal women, this lyzozeme is abscent because of the atrophy of lacrimal gland, the site of its production[43].

The explanation of the cause of night - blindness in individuals deprived of vitamin A is one of the clear - cut examples of metabolic etiology of ocular disease[44,45]. Vitamin A is oxidized by ADH and coeuzyme I to the aldehyde form called retinene, this disease is a result from lack of substrate for the enzyme available.

B - Isoenzymes in Medicine

Certain enzymes may exist in a variety of hybridized forms called isoenzymes, these isoenzymes retain their substrate - specificity, but show differences in their physical, chemical and kinetic properties, one of these enzymes which is known to occur in multiple molecular forms is the enzyme LDH, a tretramer enzyme contain all possible combinations of two different polypeptide chains known as H and M as an indication for heart and muscle tissues respectively, the enzyme LDH has five isoenzymes from LDH$_1$

to LDH_5, and the LDH isoenzyme pattern is modified in various
disease states. The clinical value is due to the fact
that the changes are often characteristic of a particular
disease and of the stage of that disease[46].

C - Treatment of Diseases by Enzymes

One of the most interesting fields in enzymatic
studies, which has received a very considerable amount
of attention has been, the use of enzymes as a pharmacologic
agents, this field has come more and more into use.

The proteolytic enzymes particularly TR, are rec-
eiving attention in the field of enzyme therapy. Using
purified intravenous TR, Innerfield - Angrist, and Schwartz[47]
reported subsidence of acute inflammation in local sites,
they described the parenteral use of TR in 588 patients
with acute inflammatory reactions, and found that there
was a rapid suppression of acute inflammation of whatever
cause. Purified TR was used for the treatment of endo-
bronchial disease in 1952. In eye surgery, relief of
edema occured in 1 - 2 days by the use of TR[48, 49].

Hyaluronidase[50] has been used with some success in
enhancing the absorption of blood from joint cavities in
hemophilia and other hemorrhagic diseases. Hyaluronidase
may also aid in wound healing.

Pancreatic DNase[51] in addition to its effect in clearing the respiratory tract has been reported to hasten cure of meningitis by intrathecal use and to hasten the cure of lung abscess by intravenous use.

This is only the begining of enzyme therapy, other enzymes and methods of activating and using in vivo enzyme activity, will undoubtedly be rapidly developed.

IV.   Determinations of Enzymes in Diseases

A -   Principles of Enzymatic Analysis in Medicine

Enzymes are biological catalysts serve as a very sensitive and accurate tool for determining the metabolic concentration; as the determination of glucose by GOD.

The assay of enzyme activities in organs, biological fluids, plant, drugs, culture media, ... etc., is included under the field of " enzymatic analysis "[52]. By comparison of enzymes activities for normal individuals with those of patients, it is possible to reach valuable diagnostic conclusions about that condition.

B -   Factors that Influence Enzyme Activity

Optimum experimental conditions are taken as a
basis for the determination of enzyme activity, pH, temperature, concentration of reactants, such as, substrate
and enzyme concentrations. The inhibitors and activators
influence the enzyme activity, and the choice of substrate
is of importance, since synthetic substrates are hydrolyzed
at different rates.

C -   Expression of Units in Enzymatic Analysis[53]

A unit is a quantity of the enzyme, related to
the velocity of the enzymatic reaction by an arbitary definition, so that the velocity tells one how many units of
the enzyme are present. So one unit of an enzyme may be
defined as that quantity which produces a certain rate of
reaction under certain set of defined conditions.

One unit (U) of any enzyme is that amount which will
catalyze the transformation of one micromole of substrate
per minute or, where more than one bond of each substrate
molecule is attacked, one micro - equivalent of the group
concerned per minute, under defined conditions. The temperature should be stated, and it is suggested that where
practicable it should be 30°. The other conditions, including pH and substrate concentration, should where practicable,

be optimal; where this is not so, it should be clearly stated. Where inconvenient numbers would otherwise be involved, terms such as mU, kU,.... etc., may be used.

Conventional units of enzyme activity defined by earlier investigators can in some instances be converted to I.U., providing the necessary parameters are known.

The four principle factors to be considered for conversion of conventional units are

1. Unit weight.
2. Unit volume.
3. Amount of active substrate liberated during the incubation period, t,.
4. Temperature and pH.

V — Enzyme Kinetics

A — Reaction Kinetics[53]

Enzymes kinetics deal with the study of reaction rate catalyzed by enzymes under specified conditions such as, enzyme concentration substrate concentration, pH, temperature, and others.

Velocities of reaction

The observed velocity of the enzyme reaction will be represented as v; which is the velocity expressed by the usual Michaelis – Menton type equation. When referring to the initial rate, it may be called " the initial steady – state velocity ". Capital V or the maximum value of v, corresponding to the saturation of enzymes with substrate.

Combination of enzymes with substrate

Under the equilibrium conditions, the equilibrium constant for reversible enzyme – substrate combinations, is called association constant by some authors but others use dissociation constant. The dissociation constant is the traditional method among enzymologists, deriving from the Michaelis theory, but it differs from the general convention used by physical chemists, according to this, if a combination is written thus:

A + B ⇌ AB, the equilibrium constant of this equation should be written

$$K = \frac{(AB)}{(A)(B)} \qquad (1)$$

Writting the typical enzyme reactions as

$$E + S \xrightleftharpoons{\quad\quad} ES \longrightarrow E + P \quad\quad (2)$$

Since the equilibrium between enzyme and substrate should be dealt by the association constant in the forward reaction but when we consider the products, the convention leads to the opposite result, so the products should be dealt with dissociation constant, according to this, it is inconvenience if some component of enzyme system have to be dealt with by association constant and other by dissociation constant, with respect to their combination with the enzyme. So it is decided that, all such equilibria should be treated in the same way; all equilibria involving combination of enzyme with substrate should be expressed in terms of dissociation constant rather than association constant.

Velocity constants

Small k will be used to represents a rate constant, capital K an equilibrium constant according to the following systems.

$$\text{System A:} \quad E + S \underset{k_2}{\overset{k_1}{\rightleftharpoons}} ES \underset{k_4}{\overset{k_3}{\rightleftharpoons}} EP \underset{k_6}{\overset{k_5}{\rightleftharpoons}} E + P \qquad (3)$$

$$\text{System B:} \quad E + S \underset{k-1}{\overset{k_1}{\rightleftharpoons}} ES \underset{k-2}{\overset{k_2}{\rightleftharpoons}} EP \underset{k-3}{\overset{k_3}{\rightleftharpoons}} E + P \qquad (4)$$

$$\text{System C:} \quad E + S \underset{k-1}{\overset{k+1}{\rightleftharpoons}} ES \underset{k-2}{\overset{k+2}{\rightleftharpoons}} EP \underset{k-3}{\overset{k+3}{\rightharpoonup}} E + P \qquad (5)$$

System A has been much used by enzymologists for many years past, system B has been used in enzymology during recent years, system A does not conform to the physico – chemical symbols produced by the International Union of Pure and Applied Chemistry.

The Enzyme Commission decided to use system C, this system conforms to the above recommendation of the International Union of Pure and Applied Chemistry, it is more symmetrical and will help to avoid misprints due to the dropping of signs. The enzyme commission has justified its choice since adopting system A exclusively would undoubtly save many enzymologists much trouble at the moment. To adopt system B exclusively would involve much inconvenience in changing well – established equations into new forms.

Factors influencing the kinetics of enzyme reactions

There are several factors that influence the enzyme reaction rate,

1. The effect of substrate concentration
2. The effect of enzyme.
3. The effect of temperature
4. The effect of pH.

The following section will be dealt with effect of substrate concentration, other factors are discussed in details elsewhere[54, 55, 56].

Substrate concentration influences the rate of the reaction, according to the type of substrate, the biochemical reactions are classified into the following

1. Single Substrate Reactions.
2. Bi-substrate Reactions.
3. Tri-substrate Reactions.

1. Single Substrate Reaction[57]

For a reaction

$$E + S \underset{k-1}{\overset{k+1}{\rightleftharpoons}} ES \xrightarrow{k+2} E + P \qquad (6)$$

The Michaelis curve is obtained when one plots the initial
velocity v of the reaction against the substrate concentration (S), described by an equation of the form:

$$v = \frac{V(S)}{K_m + (S)} \qquad (7)$$

which is called the Michaelis - Menten equation.

Michaelis and Menten assumed that the ES complex,
exist in rapid equilibrium with the enzyme and the substrate, and also assumed that k+2 << k-1.

V is the maximum velocity obtained when the (S) is suff-
iciently high to form ES complex, under these conditions

$$v = V = k+2 \ e \qquad (8)$$

e = total enzyme concentration

if (S) = K_m

Then

$$v = \frac{V}{(S) / (S) + 1} = \frac{V}{2} \qquad (9)$$

So K_m may be defined as the substrate concentration which
gives half - maximum velocity. At steady - state conditions

$$K_m = \frac{k-1 + k+2}{k+1} \qquad (10)$$

If the rate of breakdown of the ES complex to give products may be so slow that this complex remains essentially at equilibrium with the free enzyme and substrate, equation (10) will simplify to

$$Km = \frac{k-1}{k+1} = Ks \qquad (11)$$

In which Km is equal to the dissociation constant for the ES complex (Ks) and is therefore a measure of the affinity of the enzyme for its substrate; the lower the Ks the higher the affinity.

2.   Bi-substrate Reactions

The biochemical reactions, in vivo, are usually requiring two types of substrate inorder to be catalysed by certain enzymes. The relationship between these two substrates and the enzyme decide the type of the reaction. The following are some of these types of reactions.

a -   Compulsory order mechanism

$$E \xrightarrow{Ax} EAx \xrightarrow{B} EAxB \rightleftharpoons EABx \xrightarrow{Bx} EA \xrightarrow{A} E \qquad (12)$$

where Ax could be ATP, NADH, acetyl-coA,...etc.

b -   Random order mechanism

$$
\begin{array}{c}
\text{(13)}
\end{array}
$$

In the case of enzymes that obey Michaelis - kinetics,
i.e. reactions in which a plot of v against (Ax)
(or (B)) at a fixed concentration of the other substrate will give rise to a rectangular hyperbola, the
kinetic equation for, the reaction must take a form
similar to that of equation (7). For the majority of
two substrate reactions that obey Michaelis kinetics this
equation is as follows

$$
v = \cfrac{V}{1 + \cfrac{Km^{Ax}}{(Ax)} + \cfrac{Km^{B}}{(B)} + \cfrac{Ks^{Ax} \cdot Km^{B}}{(Ax)(B)}} \tag{14}
$$

This equation contains a Michaelis constant for each substrate (Km^{Ax} and Km^{B}) together with a combined constant
($Ks^{Ax} \cdot Km^{B}$). At fixed concentration of B, the equation takes the form

$$v = \cfrac{\cfrac{V(B)}{Km^B + (B)}}{1 + \left(\cfrac{Ks^{Ax} \cdot Km^B + Km^{Ax}(B)}{Km^B + (B)} \right) \cfrac{1}{(Ax)}} \tag{15}$$

at very large concentration of B equation (14) will simplify to:

$$v = \cfrac{V}{1 + \cfrac{Km^{Ax}}{(Ax)}} \tag{16}$$

a similar equation obtained at saturating concentrations of Ax. Equation (16) is identical to the simple Michaelis equation, and hence Km^{Ax} can be defined as the concentration of Ax which will give half-maximum velocity at saturating concentrations of B, and similarly Km^B can be defined.

The constants in equation (14) may be determined by a reciprocal method analoguus to the Lineweaver–Burk method. If reciprocals are taken the equation can be rearranged to give

$$\frac{1}{v} = \left(\frac{Km^{Ax} + \dfrac{Ks^{Ax} \cdot Km^{B}}{(B)}}{V} \right) \frac{1}{(Ax)} + \left(\frac{1 + \dfrac{Km^{B}}{(B)}}{V} \right) \quad (17)$$

A reciprocal plot of $1/v$ against $1/(Ax)$ at a series of
concentrations of B will give a series of straight lines
which may intersect above, on or below the $- 1/(Ax)$ axis.

3. Tri-substrate Reactions

The kinetic analysis is, in many way, similar to
the two - substrate systems. There are many possible
variations on mechanism for three - substrate reactions.
These can be " hybrid " mechanisms in which, e.g. two of
the substrate bind randomly and the third binds in a
compulsory order, or two of the substrates are involved
in a double - displacement type of mechanism.

The general kinetic equation obeyed by three - substrate reactions can be written as:

$$v = \frac{V}{1 + \dfrac{Km^{A}}{(A)} + \dfrac{Km^{B}}{(B)} + \dfrac{Km^{C}}{(C)} + \dfrac{K^{AB}}{(A)(B)} + \dfrac{K^{AC}}{(A)(C)} + \dfrac{K^{BC}}{(B)(C)} + \dfrac{K^{ABC}}{(A)(B)(C)}}$$

$$(18)$$

A, B, C, represents the three – substrates Km^A, Km^B and Km^C represents the three Michaelis constants for these substrates and K^{AB}, ... etc., represent multiples of constants.

B – Kinetic Parameters in Normal and Pathological Conditions

Since enzymes are responsible for catalyzing various biochemical reactions, measurement of different parameters of enzymatic reactions give valuable, informations, concerning the enzyme itself and the disease. When the enzyme is playing a big role in the metabolic pathway, this would centrized the general characteristics of the diagnosis and treatment. Many reports in the literature have focused the attention of many investigators, to the differences in the parameters of the same enzyme in pathological and normal state . The significance of these observations is not yet clear because of the undirected approach of these investigation. To get more benefits and information from the approach of kinetics, we suggest using one enzyme and then give a detailed picture of its kinetics in both pathological and normal state.

The aim of this work is to investigate the kinetic behavior of 5-nucleotidase in serum of liver cirrhosis, in order to clarify the picture of this significant enzyme.

VI -- Literature Review

A -- $5'$ - Nucleotidase ($5'$ - Ribonucleotide Phosphohydrolase EC 3.1.3.5)

$5'$ - nucleotidase catalyse the conversion of adeno-sine - $5'$ - monophosphoric acid to adenosine and inorganic phosphate, playes important role in nucleic acid metabolism through the breakdown of nucleotides.

1 -- Distribution

Because of the significance of $5'$ - nucleotidase it is distributed widely in different organisms. Many investigators have shown this distribution through their work.

A -- Bacterial and Yeast $5'$ - Nucleotidase

Kohn and Reis [58] have presented the evidence that the extracts of many species of bacteria .. Proteus, Hemophilus, Staphylococcus, Escherichia, and Clostridium , were capable of hydrolyzing both ribonudeoside $3'$ - and $5'$ - monophosphates. From many studies on the general characteristics of the enzyme, they concluded that bacterial $3'$ and $5'$ .. nucleotidases were distinct enzyme.

Neu and Heppel[59] found that the $\bar{5}$ – nucleotidase of E. coli was released into solution when spheroplasts were prepared with EDTA – lysozyme [60]. Furthermore, Harold and Neu, have studied $\bar{5}$ – nucleotidase of Aerobacter aerogenes, and Shigella sonnei[61]. The properties of a $\bar{5}$ – nucleotidase from Neurospora crassa have been studied by Ann C. Olson and Murraly J. Fraser in 1974[62], where two enzyme of $\bar{5}$ – nucleotidases were separated by poly-acrylamide gel electrophoresis[63].

B –   Snake Venom $\overset{\cdot\cdot}{5}$ – Nucleotidase

Various snake venoms contain $\bar{5}$ – nucleotidase activity, until recently only relatively Gude preparations have been available[64]. Sulkowski et al. have purified the enzyme from Bothrops atrox venom[65].

C –   Bull Seminal Plasma $\bar{5}$ – Nucleotidase

Mann in 1945[66], found that bull seminal plasma tested at pH 7 hydrolysed A – 5 – MP, Heppel and Hilmoe[67], purified $\bar{5}$ Nucleotidase from bull seminal plasma, suppo-rted by Bodansky and Schwartz[68]. Pilcher and Scote[69] have studied kinetically bull seminal $\bar{5}$ – nucleotidase.

D —   Liver $\bar{5}$ – Nucleotidase

$\bar{5}$ – Nucleotidase have been isolated from livers of many species and its activity has been identified in lysosomes, cytoplasmic supernatants and plasma membrane preparations [70], [71]. $\breve{5}$ – Nucleotidase from rat liver has been purified by Arsenis and Touster[72], the $\bar{5}$ – Nucleotidase of chicken liver has been partially purified from acetone powder preparations[73]. A similar enzyme preparation has been obtained from rat, frog, and pig liver acetone powder preparation[74]. Fritzon[75], [76] has provided evidence that two $\bar{5}$ – nucleotidase exist in his studies on soluble fraction of rat liver. $\bar{5}$ – Nucleotidase was also partially purified from the supernatant fraction of rat liver[77].

Distribution of $\breve{5}$ – nucleotidase between nuclear and the microsomal fraction suggested the presence of two species of this enzyme in rat liver and the possible difference in the properties of isoenzymes has been studied[78], on the other hand, some evidence have been presented that rat liver nuclei when isolated from cytoplasmic contaminant contain very little $\bar{5}$ – nucleotidase . Widnell and Unkeless[79] have obtained a highly purified $\bar{5}$ – nucleotidase from rat liver microsomes and plasma membranes using classic fractionation procedures in the presence of detergent.

E —   Intestinal $\bar{5}$ – Nucleotidase

Center and Behal[80] have isolated $\bar{5}$ – Nucleotidase from calf intestinal mucosa, Burger and Lowenstein, in 1970[81] studied the enzyme $\ddot{5}$ – nucleotidase from smooth muscle of small intestine.

F — $\ddot{5}$ – Nucleotidase from Nerve Tissue

Reis,[82] found $\bar{5}$ ·· Nucleotidase activity in the brain of the rabbit, and rat, in 1937, showed the presence of the same enzyme in calf brain and hourse nerve.

An enzyme activity in central nervous system was originally found in crude homogenate and appear to bo distinct form AP and AcP[83], $\bar{5}$ – nucleotidase has been partially purified from the supernatant fluid of sheep brain homogenates by Ipata [84, 85, 86].

G — $\ddot{5}$ – Nucleotidase from Cardiac Tissues

The myocardium's $\bar{5}$ ·· nucleotidase has recently attracted attention because of the possibility that adenosine is a physiological regulator of coronary blood flow[87].

Histochemical evidence has indicated that the enzyme resides in the sarcoplasmic reticulum and transverse tubular system of rat myocardium[88]. Other evidence indicated that enzymic activity of $\bar{5}$ - nucleotidase is also localized within the walls of the coronary blood vessels[89, 90].

II - $\bar{5}$ - Nucleotidase from Pituitary Gland

Very limited reports on the pituitary glands preparation containing $\bar{5}$ - nucleotidase activity have been described, Lisowski[91, 92], had partially purified the enzyme from this gland.

I - $\bar{5}$ - Nucleotidase from other Tissues

$\bar{5}$ - nucleotidase present in the supernatant fraction of rat and guinea pig skeletal muscle extracts has been studied[93], the enzyme in mouse kidney has been examined histochemically and electrophoretically and found to exist as isoenzyme[94]. The enzyme exists also as isoenzymes in many other tissues of mouse such as liver, spleen, intestine, tests, and heart [95], by electrophoretic techniques.

J -- $\overline{5}$ - Nucleotidase from Blood Serum

Theodore F. Dixon and Maxy Purdon[96], have demons-
trated the serum $\overline{5}$ - Nucleotidase activity by substracting
the non - specific phosphatase activity with glycerophosphate as substrate, at pH 7.5 from the total activity with
A - 5 - MP, this determination has clarified the role of
the enzyme in liver disease, the result of this work show
that a high serum AP level from liver disease is usually
accompanied by $\overline{5}$ - nucleotidase value increased as much
as 100 fold where as that from bone disease is not.

Young in 1958 [97] has shown the $\overline{5}$ - Nucleotidase
activity in human serum, in group of 30 normal adults,
and he found that its value doesn't influenced by age[98,99].

2 --   Transport

The transport of $\overline{5}$ - Nucleotidase is not clear
yet, although it is very important to know how it is
transported to serum and other tissues, knowing the transport mechanism, will give us much detailed picture on
the role of this enzyme in the metabolic pathway.

3 - Structure

The structure elucidation of $\overline{5}$ - nucleotidase has not been determined yet.

4 - Function

$\overline{5}$ - Nucleotidase is specific for nucleotides phosphorylated on the fifth ribose carbon atom[100 - 103], yeilding a nucleoside and phosphate, by hydrolysis at an optimum pH, as shown in the following reaction:

$$A - 5 - MP \xrightarrow[H_2O]{\overline{5} - Nucleotidase} Adenosine + H_3PO_4$$

5 - Specificity

$\overline{5}$ - Nucleotidase splits phosphate from muscle adernylic acid as well as other nucleotides as U - $\overline{5}$ - MP, G - 5 - MP, C - 5 - MP, I - 5 - MP rapidly, but acts only slowly or doesn't work upon nucleotides which have phosphate in position 3 of the ribose group as yeast adenylic acid (A - 3 - MP).

Gullond and Jackson[104] have found that these differences appears with fresh enzyme, but the dialyzed enzyme shows no such specificity of action. Bacterial $\overline{5}$-nucleotidase[105] catalyzes the hydrolysis of all $\bar{5}$ - ribose and $\bar{5}$ - deoxy-ribo nucleotides with preferance for A - 5 - MP. It doesn't attack A - 2 - MP, A - 3 - MP, A - 2 : 3 - MP or inorganic pyrophosphate. The snak venom $\bar{5}$ - Nucleotidase[105] hydrolysis all ribo and deoxy ribo - $\bar{5}$ - nucelotides, with greatest reactivity to A - 5 - MP, it doesn't attack $\bar{3}$ - nucleotides ATP ribose - 5 - phosphate , inorganic phosphate, or p - nitrophenylphosphate)

Liver $\bar{5}$ - nucleotidase[105] is unusual in that it hydrolyses $\bar{2}$ - , $\bar{3}$ - , and $\bar{5}$ - mononucleotides equally well with preference for d - A - 5 - MP, it also hydrolyses FMN, p - nitrophenylphosphate, and B glycerophosphate, but not inorganic pyrophosphate or bis (p - nitrophenyl) phosphate.

The enzyme from intestinal[105], hydrolyses all $\bar{5}$ - ribonucleotide at similar rate and hydrolyses $\bar{5}$ - deoxy-ribonucleotide more slowly. These properties indicate that it is similar to the one obtained from aceton powder preparation from chicken and rat liver and from soluble separation from rat liver.

The specificity of $\bar{5}$ - nucleotidase from pituitary gland[105] seems different from that of other tissues in that G - 5 - MP, U - 5 - MP are the preferred substrates. It is reported that $\bar{5}$ - nucleotidase from nerve tissue to be specific for A - 5- MP and I - 5 - MP, A - 2 - and A - 3-MP are not hydrolyzed.

6 -   Kinetic Parameters

The effect of Metal Ions

The effect of metal ions appear to vary with the
source of enzyme, according to previous studies by
Reis[106, 107], magnesium activates this enzyme to a much
smaller extent than it does to the non - specific phos-
phomonoesterase, Kay[108] in 1955 mentioned that zinc
inhibits $\bar{5}$ - nucleotidase. But Ahmed and Reis in 1958[109]
using human tissues extracts found that magnesium ions
have little effect, while manganese ions increased the
activity of $\bar{5}$ - nucleotidase by 60%. A very strong inhibition was obtained with nickel and zinc ions.

pH Optimum

The pH optimum of $\bar{5}$ - nucleotidase depends on the
source of the enzyme. Hardonk[110] presented evidence
that one $\bar{5}$ - nucleotidase seem to show greatest activity
at pH 5.0, while the other is most active at pH 7.0 - 7.5.

The purified enzyme from bull seminal plasma had
a pH optimum at 8.5[67], Song and Bodansky[111], found that
there is only one optimum at pH about 7.5 of $\bar{5}$ - nucleotidase from rat liver plasma membrane, in the absence of

magnesium ions. Sulkowski et al.[65] found that the optimum pH of $\bar{5}$ - nucleotidase from venom of Bothrops atrox, lies close to pH 9.0.

Levin and Bodansky[112] found double pH optimn for $\bar{5}$ - nucleotidase, from bull semen, namely pH 8.5 and pH 9.1 - 9.3, the second pH optimum was independent of buffer, but was magnesium ions and temperature dependent, it was also dependent upon the nature of the substrate.

Burger and Lowenstein[81] found that the enzyme $\bar{5}$ - nucleotidase possess three pH optima, the lowest of these optima (5.5) was observed only when the enzyme is not inhibited by buffer anions, the highest (9.4) is observed only in the presence of magnesium ions.

Inhibitors

A specific protein inhibitor for $\bar{5}$ - nucleotidase has been purified from E. coli, cell cytoplasm[114], it also inhibits the hydrolysis of A - 5 - MP by the $\bar{5}$ - nucleotidase from A. aerogenes, S. sonnei, and S. typhimurium, the relevance of this inhibitor protein to the action of the enzyme in vivo is not known.

The enzyme from sheep brain homogenates is strongly inhibited by very low concentration of ATP, UTP, and CTP but not by GTP, A - 2 - MP, A - 3 - MP[85].

ATP and TTP were competitive inhibitors of A - s - MP hydrolysis by $\bar{5}$ - nucleotidase from Ehrlich ascites ·· tumour cell[115]. $\bar{5}$ - Nucleotidase prepared from smooth muscle is strongly inhibited by ADP, this inhibition is much stronger than the previously described inhibition of $\bar{5}$ ·· nucleotidase by ATP, the $\propto,\beta$ - methylenephosphonate analogue of ADP is an exceptionally powerful inhibitor of the enzyme[81].

Using a partially purified enzyme from bull seminal plasma, Bodansky, and Schwartz[68] found that in the presence of magnesium ions. L - histidine inhibited the enzyme below pH 7.5 but activated it above this pH value, shifting the optimum from 7.5 to 9.3, in the absence of magnesium ions, L - histidine produced inhibition below pH 9.

Michaelis - Menten constants of $\ddot{5}$ - nucleotidase

The Km values for yeast $\bar{5}$ - nucleotidase[116] for A - $\bar{5}$ - MP, G - 5 - MP, I - 5 - MP, C - 5 - MP, and U - 5 - MP were $<$ 0.2, 0.35, 0.54, 2.55, and 2.05 x 10^{-3} M.

The Km value for $\bar{5}$ - nucleotidase from yeast are
in the range of 2 mM for purine nucleotide.

The enzyme from human serum has a Km of 0.2 mM.
The Km value for brain microsome $\bar{5}$ - nucleotidase was
2×10^{-4} M and 4×10^{-4} M for G - 5 - MP[117].

Activity Measurements

1. Colorimetric methods are usually used for the
    determination of $\bar{5}$ - nucleotidase activity[78, 99, 118]
    Earlier procedures by Reis[83] and Purdom[95], measured
    non - specific phosphotase using phenyl phosphate
    or B - glycerophosphate as substrate in duplicate
    with Adenosine - 5 - phosphatase activity at pH
    7.5, and by substraction obtained a $\bar{5}$ - nucleotidase
    activity. Other methods relied upon selective
    inhibition. Young[97] preincubated serum with EDTA,
    which only inactivates non - specific phosphatase,
    and Campbell[99] used nickel which inhibits $\bar{5}$ - nucleo-
    tidase only.

2. Spectrophotometric Methods

A method was described by Ellis[119] for the continous
spectrophotometric determination of $\bar{5}$ - nucleotidase

activity in human serum by using A - s - MP as
substrate, and coupling the nucleotidase reaction
with the ADA, and the GLDH reaction, as shown in
the following linked reactions

$$A - 5 - MP \xrightarrow[\text{Mg}^{+2}]{\text{5 - nucleotidase}} \text{Adenosine} + P_i$$

$$\text{Adenosine} \xrightarrow{\text{ADA}} \text{Inosine} + NH_4^+$$

$$NH_4^+ + 2 - OX_oG \dashrightarrow[\text{GLDH}]{} L - \text{Glutamate}$$

$$\swarrow \qquad \searrow$$

$$\text{NADH} \qquad NAD^+$$

The overall reaction was monitored by the fall
in A_{340}.

3. Determination of NH_3

A method for determination of serum nucleotidase
was described in which serum nucleotidase hydrolysis
A - 5 - MP to adenosine and inorganic phosphate,
acitvity of 5 - nucleotidase is measured by the
amount of NH_3, liberated from adenosine after
incubation with an excess of ADA[120].

Other methods were used for the determination of
$\bar{5}$ - nucleotidase activity as the use of uridine - 5 -
fluorophosphate as substrate for $\bar{5}$ - nucleotidase[121]

7 -   Clinical Interpretations of $\bar{5}$ - Nucleotidase

Although $\bar{5}$ - nucleotidase is widely distributed
in human tissues, an increase in the serum level of the
enzyme appears to occur only in liver disease. Dixon
and Purdom[96], using serum for assays of $\ddot{5}$ - nucleotidase
and AP in a variety of disease states, found that, the
group which contained, Rheumatoid arithritis, pulmonary
tuberculosis, muscle injury and others, exhibited normal levels of AP and $\ddot{5}$ - nucleotidase in the second
group $\bar{5}$ - nucleotidase was normal and non - specific AP
value were high, this group contained, rickets, pagets
disease s, Osteogenic tumours, breast cancer, Albrights
syndrome, and spinal neoplasm. In the third group comp-
rising patients with jaundice, tuberculosis, and amyloi-
dosis and one case of polycystic kidney with uremia.

Young[97] in 1958 in a study of serum AP and $\bar{5}$ -
nucleotidase in hepatic disorders found that 67% of
patients with biliary tract obstruction had serum activities greater than 6 times the upper limit of normal.

In hepatogenous jaundice on the other hand 31% showed normal activity, a further 59% were less than three times and non was greater than six times the upper limit of normal.

Serial estimation of serum AP, and $\bar{5}$ - nucleotidase levels have been made in patients with acute and chronic liver disease, little difference was observed in the response of the two enzymes in acute infective hepatitis, however in chronic disease (including chronic active hepatitis and primary biliary cirhosis) serum $\bar{5}$ - nucleotidase was more markedly and persistently elevated than serum AP[122]. It was also found that the activity of $\bar{5}$ - nucleotidase in serum, in patients with early, cirrhosis and hepatic coma, is increased (as compared with normal controls). But a statistically significant decrease was observed in advanced cirrhosis.[123]

Serum $\bar{5}$ - nucleotidase has been determined in control groups and in patients with various types of neoplastic diseases. The $\bar{5}$ - nucleotidase activity is elevated in cancer patients with hepatobiliary disease .

B -   Liver Cirrhosis

Hepetic cirrhosis is a disease characterized by
chronic, diffuse inflamation of the liver, accompanied
by proliferation of connective tissue, degenerative and
regenerative changes in liver cells, resulting in dis-
organization of lobular architecture.

Liver cirrhosis is classified separately according
to morphological, etiological and functional properties[124].

Morphological Properties[125]

Three anatomical types are recognized

a. Micronodular, is also called " portal " or
   " Septal ", may represent impaired capacity for
   regeneration as in malnutrition, old age...etc.

b. Macronodular, is also called " post - necrotic"
   following liver - cell necrosis.

c. Mixed type of cirrhosis, the liver shows both
   micro and macronodular features.

Etiological Properties

a. Viral hepatitis, this disease with characteristic liver lesions is caused by more than on virus. The term virus hepatitis includes infectious (infective) hepatitis, and serum hepatitis. It is not clear whether the etiologic agents that induce these two patterns are identical, different strains of the same virus or two quite separate -- closely related viruses[126].

b. Alcoholism

The principal effect of alcohol is by distortion of the normal food intake, to produce protein and vitamin deficiencies leading to fatty infilteration of the liver. But some of the alcohol effect is probably due to mild prolonged toxic action of liver cells.

c. Malnutrition

Malnutrition leads to sever liver damage, which, in a pregnant mother, may be reflected in the off spring by such condition as Kwashiorker accompanied by sever anemia[127]. Hepatic necrosis and fibrosis can be produced in experimental animals by appropriate diets particularly those low in protein and essential amino acids[125, 128].

d. Chemical Poisoning

Necrosis of the liver may be caused by drugs, poisons such as (phosphorus, mercury), and by substances used in technical and manufacturing processes (carbon tetrachloride, tetrachloroethane and trinitro toluene). The action of these substances can be observed both in man and in experimental animals. The effect of these substances on the development of liver damage depends both on size of dose and the length of time the poison act.

e. Cardiac cirrhosis (Cardiac Sclerosis)

The clinical recognition of cardiac cirrhosis is usually impossible[130], the spleen is not larger than in those in heart failure but without cirrhosis, other features of portal hypertension are usually absent except in very severe cardiac cirrhosis.

f. Syphilitic Cirrhosis

In both congenital and acquired syphilis, cirrhosis of the liver with many of the signs and characteristic of the whole group of cirrhosis may be found. In areas where syphilis is well controlled and treated, this is becoming a very rare type of cirrhosis.

3. Cryptogenic Cirrhosis

The patients with this kind of cirrhosis do not give a history of hepatitis and have severe been jaundiced. The pathology is that of macronodular cirrhosis.

Functional Properties

Liver cirrhosis is also classified according to the functional properties, the liver failure is estimated by such features of Jaundice, ascites, pre-coma, coma, low serum albumin and raised transaminase levels and a pro-thrombin deficency not corrected by Vitamin K.

So in every patient – diagnosis must be in terms of morphology etiology and hepatic function.

Symptoms and Signs of Cirrhosis

All forms of cirrhosis, when severe, have certain clinical symptoms and signs in common. These stem from the hepatic cell destruction, regeneration and fibrous scarring characteristic of all forms of cirrhosis.

The destruction, regeneration and scarring in turn lead to portal hypertension, which then serves as an additional mechanism in the production of the clinical syndrome of cirrhosis. Each of the major clinical features will be discussed separately:

1- Portal hypertension

Portal hypertension refers to an increase in blood pressure within the portal vein from a normal of 50 – 100 mm. of water up to 50C mm. or higher[126].

48

It is oversimplification to ascribe intrahepatic portal hypertension to obstruction of the radicales of the portal vein by fibrous scarring alone. Although this is undoubtedly the single most important factor, there are other influences that contribute to obstruction and raised pressures such as swelling of the liver cells, regenerative changes with resultant distortion nodules.

2 -   Ascites

Ascites may be defined as an excess accumulation of transudate within the peritoneal cavity. Ascites fluid contains 1 - 2 grams of albumin per 100 ml. of transudate, and only a small number of cells, principally lymphocytes and mesothelial cells.

This fluid usually occurs late in the cirrhosis but may at times be the presenting complaints. The mechanisms by which cirrhotic ascites is produced, depends on many systemic and local factors.

The systemic factors[126], include low serum proteins (albumin) due to impaired synthesis of albumin by the liver and increased aldosterone and antidiuretic hormone activity with retention of sodium and water. The low serum albumin causes the development of ascites when the hydrostatic pressure in the portal vein exceeds the colloid pressure of the blood.

The local factors, include portal hypertension, which
is considered the principal local factor which cause the
accumulation of the fluid, in the peritoneal cavity.

In some patients the ascites fluid accumulation
remains moderate and do not require mechanical removal,
but in most cases the swelling of the abdomin becomes into
lerable, so paracentosis must be performed to give relief.

3 -   Splenomegaly

Congestive splenomegaly, with weight -- increases
up to 1000 grams, is a frequent feature of cirrhosis. As
a consequence of the splenic enlargement, the " hypersplenic
syndrome " may develop a wide variety of hematologic abnor-
malities, including anemia, leukopenia and others.

4 -   Hepatomegaly

The liver is enlarged in most cases of cirrhosis,
this is especially true with fatty nutritional cirrhosis,
in which liver enlargement can be demonstrated, in about
three fourth of the cases, hepatomegaly may also be present
in certain forms of biliary cirrhosis[126].

5 -   Gastrointestinal bleeding and portal vein by-passes

As the portal hypertension develops, a large volume
of blood is rerouted from the portal vein to the systemic
system without passing through the liver. The principal
collateral channels involved are the esophageal venous
plexus, the hemorrhoidal plexus and the subcutaneous veins
of the abdominal wall. The frequency of esophageal varices
with cirrhosis depends to a considerable degree on the type
of the cirrhosis and the techniques used in diagnosing the
varices. The enhancement of circulating fibrinolytic activity in patients with cirrhosis of the liver encourages
bleeding from esophageal varices, on the other hand the
inability of the liver to synthesize fibrinogen at an
optimal rate further weakens the hemostatic mechanism[131].

Drugs capable of enhancing thrombolytic activity
can evoke exaggerated bleeding responses in cirrhotic
patients[132].

6 -   Anemia

Anemia is common in some cases of liver cirrhosis
as in Laennec's cirrhosis. The degree of anemia bears
only a poor correlation with the severity of the disease[133].

The characteristic type is either macrocytic or normocytic, the red blood cells showing little variation in size and shape.

The anemia does not respond to the administration of folic acid or liver extract but shows gradual improvement with recovery from the underlying disease. Schiff[134] and his associates demonstrated the presence of antianemic substances (against pernicious anemia) in livers of patients who died from hepatic cirrhosis with macrocytic anemia. Therefore, the anemia does not appear to be due to lack of storage of antianemic principle. So the mechanism responsible for the anemia of cirrhosis is not understood.

7 -   Endocrine disturbances in cirrhosis

The steroid hormones of the adrenals and gonads, are metabolized largely in the liver. Therefore when cirrhosis renders the liver unable to continue this function, a large variety of endocrine derangements follow. These derangements may be loss of potency, and infertility, all of which are attributed to an increase in circulating estrogens levels, and the failure of the damaged liver to detoxify it.

8 -   Abdominal pain

Abdominal pain is present in varying degree.   It
has no peculiar attributes in cirrhosis, it may be mild,
dull, sharp, wavelike, or steady, this abdominal pain may
be confined to the region of the liver or referred to the
epigastrium or the lower abdomin.   This symptom has been
attributed to capsular swelling, perihepatitis, spasm of
the biliary ducts, or to intermittent vascular spasm[133].

The Principles of Treatment of Cirrhosis

The most important principals in the treatment of
cirrhosis are [124, 135]:

1 --   Rest in bed.
2 --   A good diet rich in protien and vitamins, low in fat.
3 --   Abstinence from alcohol.
4 --   Choline and methionine have been widely used in the
treatment of cirrhosis.
5 --   For ascites treatment, diuretics are given.   A thiazide is the choice for initial use, but frusemide
(40 - 80 mg) is used latter because it is more
potent than the thiazide, and has a rapid action.

6 - Prednisolone increases appetite and well - being, while serum globulin, transaminases and bilirubin values fall.

7 - Corticosteroid drugs inhibit immune and inflammatory mechanisms.

8 - Blood transfusion is usually necessary in the emergency treatment, and the amount required may be very large.

Biochemistry of Liver Cirrhosis

Various biochemical constituents are effected in different cases of liver cirrhosis, certain facts concerning the serum proteins are of importance in connection with the interpretation of the diagnostic significance of changes that occur in diseases of the liver, and bile passages. Hypoalbuminemia occur in chronic liver diseases [124, 133, 136], as liver cirrhosis, the albumin concentration is depressed usually between 2.0 and 3.5 gm/100 ml[135]. The cause of this hypoalbuminemia is explained by the fact that, in normal persons albumin is mainly synthesized in the liver cells, so the presence of severe liver damage as in cirrhosis, causes a decrease in albumin synthesis and then causing hypoalbuminemia. Accompanying low albumin concentration is an elevation in globulin concentration caused principally by a rise in gammaglobulin portion [135], the gammaglobulin concentration lies between 3.0 and 4.0 gm/100 ml in liver cirrhosis, this rise in gammaglobulin is attributed to the increase in Kupffer cells, but this explanation is inadequate.

The alphaglobulins may be decreased in liver cirrhosis but are usually within normal limits, changes in Betaglobulins are encountered more consistently than those in Alphaglobulin, they may be increased in liver cirrhosis.

The serum total cholestrol concentration may be normal in cirrhosis, but the ester concentration is usually depressed[133, 135]. As the disease advances, the total concentration tends to diminish because of the fall in cholestrol esters. There is also a slight increase in serum bilirubin level[133], in cirrhotic patients.

In urine urobilinogen is present in excess, in cirrhotic patients, it is produced mainly by bacterial action on bilirubin in the small intestine, it is partially reabsorbed into the portal circulation being re - excreted by the liver in the bile, the remainder is changed further to steroobilin.

There is evidence that the out put of antidiuretic material (Vasopressin) in urine is increased in patients with hepatic cirrhosis[137], in such cases, the diver may have lost its normal capacity to inactivate the hormone[131].

Enzymatic Changes In Liver Cirrhosis

It has been demonstrated that, different biochemical reactions involved in the metabolism are enzyme catalyzed, and since many enzymes of cell metabolism are present in serum of normal individuals any alteration of the serum enzyme activity is an indication of abnormal biochemical events in the tissue of origin.

In liver cirrhosis, reports indicated that a rise in the activity of AP, GPT, GOT, γ - GT, $\bar{5}$ - nucleotidase, may be observed in at least one stage of the disease. The major value of $\bar{5}$ - nucleotidase assay is its specificity for hepatobiliary diseases. In primary and secondary biliary cirrhosis, the $\bar{5}$ - nucleotidase activity rises to high values, depending on the intra or extra - hepatic of cholestasis.

The above mentioned reports indicate that very limited amount of work has been done on serum $\bar{5}$ - nucleotidase in liver cirrhosis, so it was intended to study the characteristics of $\bar{5}$ - nucleotidase in liver cirrhosis in comparison to normal subjects to find out whether the changes brought about by the liver cirrhotic transformation are of a quantitative or qualitative nature, through kinetic studies.

CHAPTER TWO
EXPERIMENTAL

E X P E R I M E N T A L

I — Materials

A — Chemicals

A - 5 - MP, 2 - d - A - 5 - MP (as sodium salts) were purchased from Sigma Chemical Co., ATP (as sodium salts), adenosine were from BDH Chemical Co.,

Veronal buffer constituents and all other chemicals used were of analytical grade reagents.

Water distilled from glass was used throughout the experiments.

B — Specimen

A total of 63 normal blood samples of both sexes and different ages were obtained from healthy students and staff members of the College of Science, by venipuncture. These samples were used both for enzyme activity determination and kinetic measurements.

52 blood samples of both sexes and different ages from untreated patients with liver cirrhosis who were

57

admitted to the Medical City, Al - Yarmouk Hospital, and All- Shaab Hospital. Diagnosis by specialists were based on patients complete histories, physical examination, and liver - biopsy. patients having a history of diseases other than liver cirrhosis were not included in this study.

Blood samples were usually left at room temperature for about one hour, the sera were then separated by spinning for 5 - 10 minutes at 3500 r.p.m., at room temperature.

Analysis on normal and liver cirrhotic sera were always performed on the same day of sample collection.

II - Instrumentation

The Coleman Model 6/20 Junior II Spectrophotometer was used for absorbance measurements, it is equipped with cuvettes for sample or reference material, its wave length range from 325 nm to 825 nm, the scale panel provides transmittance scale (black) and absorbance scale (red), transmittance is calibrated from 0% to 100% and absorbance is calibrated from 2.0 to 0.

JANETZI Laboratory Centrifuge Type T5 (maximum speed 5500 r.p.m.) was used for the separation of serum samples.

59

The Pye Unicam pH Meter Model PW 9418 was used for pH measurements.

Water bath, fixed at $37^{\circ}C$ was used for incubation the samples, throughout the whole of this work.

III – Methods

A –   Assay of $\bar{5}$ – nucleotidase in sera of normal indivi-
        duals and liver cirrhotic patients, at 37°
        (Campbell's method)[99]

Principle

$\bar{5}$ – nucleotidase catalyzes the dephosphorylation of nucleotides that have the phosphate group attached to C-5 of the ribose radical. At pH 7.5 there is a significant hydrolysis of the substrate by non – specific AP, for which a correction must be applied. The method involves two parallel enzyme activity determinations with A – 5 – MP as substrate. In one the presence of nickel specifically inhibits $\bar{5}$ – nucleotidase, and therefore estimates the hydrolysis of the substrate by non – specific AP. In the second, the absence of nickel allows the estimation of total phosphatase activity. The difference in activity (in terms of inorganic phosphate liberated) gives the $\bar{5}$ – nucleotidase activity.

Inorganic phosphate was estimated by the method of Fiske and Subbarow,[138] in which the inorganic phosphate couples with molybidic acid to form a yellow phospho-molybdate, this can be reduced to give a blue colour which is directly proportional to the amount of inorganic phosphate present.

A unit of enzyme activity was defined as, /u moles of inorganic phosphate produced per min. per liter of serum, which was expressed as I.U./liter.

Reagents

a. Veronal buffer pH 7.5 - 8.25 g. of sodium barbiturate (sodium - 5, 5 - diethyl barbituric acid) in 700 ml distilled water, adjusted to pH 7.5 with 2N HCl, and diluted to 1000 ml.

b. Manganese sulfate - 3.38 g. in 900 ml veronal buffer pH 7.5, (the pH was adjusted to 7.5), and the volume was completed to 1000 ml.

c. Nickel chloride - 23.77 g. in 1000 ml distilled water.

d. A - 5 - MP - 0.0499 g. in 10 ml 30 mM veronal buffer prepared freshly.

e. TCA - 10 g. in 100 ml distilled water.

61

f.   Ammonium molybdate -- 5 in 100 ml distilled water.

g.   Stock phosphate standard - 2.19 of potassium
       dihydrogen phosphate in 500 ml distilled water,
       kept at 4°C with few drops of chloroform.

h.   Working phosphate standard - 1 ml of stock standard
       in 100 ml 5% TCA, prepared freshly.

i.   Perchloric acid 60%, analytical reagent.

j.   Ascorbic acid - 50 mg in 25 ml distilled water.

Methods

Each assay system ... consisted of the following
tubes

1 -   Test - 1.5 ml of 30 mM veronal buffer pH 7.5 was
       mixed with 0.1 ml of 1 mM manganese sulfate.

2 -   Control - 1.3 ml of 30 mM veronal buffer pH 7.5
       was mixed with 0.1 ml of 1 mM manganese sulfate,
       and 0.2 ml of 10 mM nickel chloride.

To both tubes, 0.2 ml of serum was added, then they
were placed in a water bath for 3 minutes at 37°,
0.2 ml of A - 5 - MP 1 mM was added, then they
were left for exactly 30 minutes, the reaction was
then stopped by the addition of 2 ml of 10% TCA.

The tubes were shaken well, removed from the water bath and spinned for 10 minutes at 3500 r.p.m.

3 - Standard - 1 ml of working phosphate standard was added to 1 ml of 10% TCA.

4 - Blank - 1 ml of H_2O was added to 1 ml of 10% TCA.

The concentrations of substrate, activator, inhibitor and buffer, given were calculated as the final molarity in the reaction mixture.

For inorganic phosphate determination, to all tubes 0.1 ml of 60% perchloric acid   0.1 ml of ammonium molybdate, and 0.1 ml of ascorbic acid, were added, the tubes were shaken gently after each addition, and then read at 700 mu after 10 minutes.

All the reaction velocities were determined at 37°, during the initial zero order portion of the time curve.

Calculation

The standard tube contains 10 /ug of phosphate, thus the phosphate produced in the enzymic reaction by 0.1 ml of serum is equal to

$$\frac{T - C}{S - B} \times 10/u_g = \frac{T - C}{S - B} \times 10 \times \frac{1}{31} = /u \text{ Mol. of}$$

inorganic phosphate.

where

31 = MW of inorganic phosphate

Therefore 1 liter of serum in 1 minute produces

$$\frac{T - C}{S - B} \times \frac{10}{31} \times \frac{1000}{0.1} \times \frac{1}{30} = \frac{T - C}{S - B} \times 108 /u \text{ Mol of}$$

inorganic phosphate produced per min. per liter (I.U.)

B -   5 - Nucleotidase kinetic studies at 37 in normal and liver cirrhotic sera

Serum samples of untreated liver cirrhotic patients and normal controls were used throughout all kinetic studies.

All the reactions were carried out at 37°, in 30 mM veronal buffer pH 7.5, the enzyme activity was calculated as I.U./litre.

The concentrations of substrate - inhibitors, activities, were dependent on the particular group of experiments. In the experiments for the determinations of Km, optimum substrate concentration, and other parameters six to seven different substrate concentrations were employed. In those experiment concerned with the effect of pH the final concentration of A - 5 - MP was 1 mM, at a final concentration of 30 mM veronal buffer.

All reaction velocities were determined during the initial zero order portion of the time curve.

1 -   Determination of the optimal substrate concentration for 5 - nucleotidase for both normal and liver cirrhotic sera, at 37°

The same method was used as under point A, for the assay of 5 - nucleotidase, 2 ml of incubation mixture was prepared which/consisted of, 1.5 ml 30 mM veronal buffer pH 7.5, 0.1 ml 1 mM manganese sulfate, 0.2 ml serum, 0.2 ml A - 5 - MP, the concentrations of A - 5 - MP employed were 0.05 mM, 0.1 mM, 0.2 mM, 0.4 mM, 0.6 mM, 0.8 mM, 1.0 mM, the concentrations of substrates given were calculated as the final molarity in the reaction mixture, the control contained the same concentration of A – 5 – MP as in the test. These concentrations were calculated and prepared as follows:

0.2 mL of 10 mM A – 5 – MP (stock) in 2 ml reaction mixture, resulted in a final molarity of 1.0 mM. The other concentrations were prepared as in table (A) below.

Table – A

ml. of stock A – 5 – MP	ml of 30 mM veronal buffer	Final A – 5 – MP molarity
Stock, 10 mM		
1 ml	0	1 mM
0.8 ml	0.2	0.8 mM
0.6 ml	0.4	0.6 mM
0.4 ml	0.6	0.4 mM
0.2 ml	0.8	0.2 mM
0.1 ml	0.9	0.1 mM
0.05 ml	0.95	0.05 mM

2 - The effect of time on the reaction of 5 - nucleotidase in normal and liver cirrhotic patients at 37°

Serum samples of untreated liver cirrhotic patients
and normal controls were used to test the effect of different time intervals (5 - 35 minutes), on 5 - nucleotidase
reaction rate. All the reactions were carried out at 37°,
in 30 mM veronal buffer pH 7.5, the concentration of
A - 5 - MP was 1 mM. The enzyme activity was calculated
as I.U./liter, using the same formula mentioned in A, for
the assay of 5 - nucleotidase.

3 - Determination of Km values for 5 - nucleotidase in
normal and liver cirrhotic individuals.

Km (A - 5 - MP) was determined from the experiment presented in section 1, the experimental data obtained
were velocities (calculated as μ MOL of inorganic phosphate
produced per minute, per liter of serum) at different concentrations of A - 5 - MP from 0.05 mM - 1.0 mM. The data
obtained were fitted in several ways to obtain the Km values.

The following methods were used for Km determinations

a - A plot of v versus (S) according to the original
Michaelis - Menten equation.[54]

b -   The Linewaver - Burk[54] double reciprocal plot.

c -   A plot of (S)/v versus (S).[54]

d -   The direct linear plot by Eisenthal and Cornish - Bowdon.[139]

This is a new plot for analysing the results of kinetic experiments in which Michaelis - Menten equation is obeyed.

The K_m values were obtained according to the following.

Two axes were set for K_m and V, corresponding to the X and Y axes, respectively. For each observation (S, v), the points K_m = - (S), were marked on the K_m axis , and V = v on the V axis, a line was then drawn through the two points extended it into the first quadrant, then this was done for all observations (5 were sufficient), the lines, intersected at a common point, whose co - ordinates (K_m, V) provided the values of K_m and V. These estimates were marked off on the axes, then the median (i.e. the middle) value from each series was taken to be the test values for V or K_m. When there were an even number of values the median was taken as the mean of the middle two estimates. If there would exist any three lines that appeared to intersect at a common point, the point was treated

as three points rather than one in finding the mudian,
also, the common intersection of four lines was treated
as six intersections, the total number of intersections
was always $\frac{1}{2}$ n (n - 1); where n represented the number
of observations. For counting intersections, it was best
to start counting from the low end of the scale and to
stop where half of the total had been counted. The reasons
for choosing the media is the best estimate were stressed
by Cornish - Bowden and Eisenthal.[140]

4 -   Affinity of $\bar{2}$ - d - A - 5 - MP to $\bar{5}$ - nucleotidase
       in normal and liver cirrhotic persons

Using the same method under point A, for the assay
of $\bar{5}$ - nucleotidase, but different $\bar{2}$ - d - A - 5 - MP
concentrations were used, each test and control contained
one of the following concentrations of $\bar{2}$ - d - A - 5 - MP,
0.1 mM, 0.2 mM, 0.4 mM, 0.6 mM, 0.8 mM, 1.0 mM. The reaction velocities were determined at 37° , in 30 mM veronal buffer

pH 7.5. The enzyme activity was expressed as I.U./liter. $\bar{2}$ - d - A - 5 - MP was prepared freshly in veronal buffer and the pH was adjusted to 7.5.

5 - **The effect of temperature of incubation on $\bar{5}$ - nucleotidase activity, in normal and liver cirrhotic sera.**

In this experiment different temperatures were used for incubation of reaction mixture for 30 minutes, these temperatures were 8° - 25° , 37° , 45° , 60° , 100° . The incubation mixture was consisted of, 1.5 ml 30 mM veronal buffer 0.1 ml 1 mM manganese sulfate, 0.2 ml of 1 mM Amp, 0.2 ml of serum.

The same method was used as under point A, except that different temperature of incubation were employed.

6 - **Stability measurements of $\bar{5}$ - nucleotidase for both normal and liver cirrhotic persons.**

One ml of serum for both normal and liver cirrhotic sera, was divided into five portions.

Portion 1   was left at room temperature for 30 min.
Portion 2   was incubated at 45° for 30 min.

Portion 3   was incubated at 49° for 30 min.

Portion 4   was incubated at 52° for 30 min.

Portion 5   was incubated at 60° for 30 min.

At the end of the incubation period, the tubes were cooled in tap water and then assayed for $\bar{5}$ - nucleotidase activity using the same method under section A.

7 -   Variation of $\bar{5}$ - nucleotidase activity, with serum concentrations, in normal individuals and patients with liver cirrhosis, at 37°.

Experiments similar to the one under point A, for the assay of $\bar{5}$ - nucleotidase, but different volumes of sera were used, the volume of the incubation mixture was kept 2 ml, by varying the volume of the veronal buffer. The volume of sera in test and control used ranged between 0.05 - 0.3 ml, diluted in 30 mM veronal buffer pH 7.5, the concentration of A - 5 - MP used was 1 mM. The velocity of the reaction was expressed as I.U./liter.

8 -   Effect of pH on v and Km values for action on
        A - 5 - MP in presence of magnesium ions, in both
        normal and liver cirrhotic sera.

        Experiments similar to that presented under point
(1) for the determination of optimal substrate concentration were performed at different pH values ranging between 7 to 8.5, the A - 5 - MP concentrations used ranged
between (0.1 - 1.2 nM) 0.1 ml 20 nM, magnesium sulfate,
(prepared freshly in veronal buffer, and pH adjusted to
7.5) was used instead of manganese sulfate in these
experiments.   The buffer used was veronal buffer.

9 -   The pH optimum of $\bar{5}$ - nucleotidase for both normal
        and liver cirrhotic sera.

        The effect of pH on the activity of $\bar{5}$ - nucleotidase
was studied by using different values of pH (7 - 8.5)
at one substrate concentration, 1 nM A - 5 - MP, veronal
buffer was used for providing these pH values.   0.1 ml
20 nM magnesium sulfate was used instead of manganese
sulfate.

10 —   Influence of metal ions on activity of $\bar{5}$ — nucleo-
        tidase, for both normal and liver cirrhotic patients. T
        effects of magnesium ions, and nickel ions were studied
        with 30 mM veronal buffer pH 7.5, for both normal and
        liver cirrhotic individuals.

a — Effect of magnesium ions

The effect of 0.1 ml 20 mM magnesium sulfate on
$\bar{5}$ — nucleotidase reaction rate, was studied in the absence
of manganese sulfate, and at varying concentrations of
A — 5 — MP (0.1 — 1.0 mM).

The same procedure was used as under point 1, the
activity of $\bar{5}$ — nucleotidase was expressed by I.U./liter.

A control experiment was run without magnesium ions,
to determine the relative reaction rates at each A — 5 — MP
concentration.

b — Effect of nickel ions

The reaction rate of $\bar{5}$ — nucleotidase was investigated in the presence of different concentrations of nickel
ions, at two substrate concentration.

The same method was used as under point A for $\bar{5}$ - nucleotidase assay, but different nickel ions concentrations were added to the test, keeping the concentration of nickel ions in the control 10 nM. The reaction mixture of the test contained one of the following concentrations of nickel chloride, 0.1 nM, 0.15 nM, 0.2 nM, 0.3 nM, 0.4 nM, these concentrations were used at two substrate concentrations 0.8 nM A - 5 - MP and 0.4 nM A - 5 - MP.

11 - Inhibition of $\bar{5}$ - nucleotidase by ATP, in normal and liver cirrhotic persons, at 37°

The reaction rate of $\bar{5}$ - nucleotidase was investigated by using different concentrations of ATP, at two substrate concentrations. The incubation mixture of the test was consisted of, 1.3 ml 30 nM veronal buffer pH 7.5, 0.1 ml of 1 nM manganese sulfate, 0.2 ml of ATP (from 0.1 nM to 0.8 nM), 0.2 ml serum; after incubation for 3 minutes, 0.2 ml of 1 nM or 0.6 nM A - 5 - MP was added. The incubation mixture of the control was composed of, 1.1 ml 30 nM veronal buffer, 0.2 ml ATP (from 0.1 nM to 0.8 nM), 0.1 ml of 1 nM manganese sulfate, 0.2 ml 10 nM nickel chloride, 0.2 ml serum, after incubation for 3 minutes, 0.2 ml of 1 nM or 0.6 nM A - 5 - MP was added. The experiment was then run as in point A. ATP was prepared freshly in veronal buffer, and the pH was adjusted to 7.5.

12 - <u>Inhibition</u> of $\overline{5}$ - nucleotidase by adenosine, for both normal and liver cirrhotic sera.

The effect of adenosine on $\overline{5}$ - nucleotidase reaction rate was investigated in the presence of two substrates concentrations, (0.6 mM and 1 mM), the incubation mixture of the test was consisted of, 1.3 ml 30 mM veronal buffer pH 7.5, 0.1 ml of 1 mM manganese sulfate, 0.2 ml of adenosine (from 0.1 - 0.8 mM), 0.2 ml serum, after incubation for 3 minute, at 37°, 0.2 ml of 1 mM or 0.6 mM A - 5 - MP was added. The incubation mixture of the control, consisted of, 1.1 ml 30 mM veronal buffer, 0.1 ml 1 mM manganese sulfate, 0.2 ml 10 mM nickel, 0.2 ml, of adenosine, (from 0.1 to 0.8 mM), 0.2 ml serum, after incubation for 3 minutes, at 37° 0.2 ml of 1 mM or 0.6 mM A - 5 - MP was added.

The experiment was then completed as under point A. Adenosine was prepared freshly in veronal buffer pH 7.5.

CHAPTER THREE

RESULTS

R E S U L T S

I – Activity measurements of $\bar{5}$– nucleotidase

Tables 1 and 2, show the activities of $\bar{5}$ – nucleotidase in both normal and liver cirrhotic sera, at 37°. The results were expressed as /u Moles of inorganic phosphate liberated / min. / liter. The analysis were performed on the same day of sample collection, as described under section " A " of the " Experimental ".

The activity for the enzyme was found to be from 5 – 50 I.U./liter, in patients with liver cirrhosis, but patients with biliary liver cirrhosis, the range of activity was 20 – 50 I.U./liter, $\bar{5}$ – nucleotidase activity for normal Iraqi individuals was measured and found to be between 1.45 – 12.09 I.U./liter, the percent increase in the activity of liver cirrhotic patients over the normal was found to be 57.37% for liver cirrhosis and 76.19% for biliary cirrhosis.

II – Stability of $\bar{5}$ – nucleotidase in both normal and liver cirrhotic individuals.

A – Crude serum of normal individuals could be stored for 20 hrs., at – 20° without loss of $\bar{5}$ – nucleotidase activity, to prevent bacterial growth it is convenient to keep the material froozen for the same period.

B –   Table 3 shows the decrease in the activity of
        $\bar{5}$ – nucleotidase which was larger for pathological
        specimen than that of normal individuals, at various
        temperatures; 27°, 45°, 49°, 52°, 60°. The activity
        was measured at pH 7.5, and 1.0 mM A – 5 – MP, as
        described in point " 6 " of the " Experimental ".

III – Effect of time of incubation on the activity of
        $\ddot{5}$ – nucleotidase for both normal controls and liver
        cirrhotic persons, at 37°, and pH 7.5 . . . .

        A – 5 – MP was converted to product at a linear
rate until hydrolysis was essentially complete (Figure
1, 2) for both normal and pathological sera. The hydrolysis of A – 5 – MP proceeded at a constant rate through
most of its course, but differs in the amount of inorganic
phosphate produced at each time intervals in both cases.
The method was described under " Experimental ".

        For normal serum, at the end of one minute 0.2%
of the substrate was used for normal serum leaving 99.8%,
during the second minute 0.2% of the remaining was used,
at the end of two minutes 0.399% of substrate was used,

and at the end of three minutes 0.598% of the substrate will be converted to product which is equal to 0.00598 mM of the substrate used.

For liver cirrhotic sera, 1.6% of the substrate was used leaving 98.4%, at the end of one minute, during the second minute 1.57% of the remaining was used, and at the end of third minute 4.7230% of the substrate will be converted to product which is equal to 0.047233 mM of the substrate used.

IV — Optimal A — 5 — MP concentration for 5 — nucleotidase for both normal and liver cirrhotic sera.

Optimal A — 5 — MP concentration was measured in the presence of 30 mM veronal buffer pH 7.5, at 37°. Different A — 5 — MP concentrations were used as explained under point " 1 " in the " Experimental ", 1 mM of A-5-MP gave optimal 5 — nucleotidase activity in both normal and liver cirrhotic sera. The activity was expressed by I.U./liter. Results obtained are shown in Figure 3.

V — Metal requirements

A — Table 4 represents the splitting of A — 5 — MP by 5 — nucleotidase, which was usually measured in the presence of 1 mM manganese sulphate or 20 mM magnesium sulfate as activators. The rate was not the same for both normal and liver cirrhotic sera.

B — In the presence of 10 mM nickel chloride the reaction was inhibited for both normal and liver cirrhotic sera, Table - 4, reveals this effect.

VI — Determination of K_m (A - 5 - MP) for 5 - nucleotidase in normal and liver cirrhotic sera.

K_m (A - 5 - MP) was determined at 37° in 30 mM veronal buffer pH 7.5, in the presence of different A-5-MP concentrations, kinetic data were plotted in various ways to determine K_m values, the direct linear plot was compared with the traditional methods currently in use as shown in Figures 4, 5, 6, 7. The results obtained by these methods, shows that the K_m of A - 5 - MP was markedly reduced in liver cirrhotic persons, this is illustrated in Table 5, which summarizes the values of K_m in different conditions. Details are as in point " 3 " " Experimental ".

VII — Activity of 5 - nucleotidase in normal and liver cirrhotic persons towards 2 - d - A - 5 - MP

A — 5 - nucleotidase activity with respect to the splitting of 2 - d - A - 5 - MP, was measured in 30 mM

veronal buffer pH 7.5 at 37°, as explained in point
" 4 " in the " Experimental " . The activity was
calculated as I.U./liter. $\bar{5}$ - nucleotidase of
liver cirrhotic persons resulted in much higher
activity than that of normal at various substrate
concentrations, 0.1 mM, 0.2 mM, 0.4 mM, 0.6 mM,
0.8 mM, 1.0 mM. The activities ratios towards
$\bar{2}$ - d - A - 5 - MP, for pathological to normal were
larger than one, using different substrate concentrations, as shown in table 6A.

B · Km ($\bar{2}$ - d - A - 5 - MP) was determined for both
normal and liver cirrhotic persons, the Km for
normal was greater than liver cirrhotic persons
(Figure 8). The experiment was run as under
" Experimental ". The Km values obtained for both
normal and liver cirrhotic individuals were greater
than that obtained for A - 5 - MP (Table 6 B).

VIII - Effect of serum concentration on the reaction rate
of $\bar{5}$ - nucleotidase for both normal and liver
cirrhotic sera.

A solution of crude serum at different values showed
a variation in the activities of $\bar{5}$ - nucleotidase expressed

in I.U./liter, when different amounts of serum were incubated with 1 mM A – 5 – MP; 50/ul, 100 /ul, 150 /ul,
200 /ul, 250 /ul, 300 /ul, as shown in Figure 9. The
activity of 5 – nucleotidase for liver cirrhotic serum
was much higher than that for normal serum, specially
at 300 /ul serum.

IX   Inhibition constants for 5 – nucleotidase in both
      normal and liver cirrhotic individuals.

A   The effect of nickel chloride on the 5 – nucleotidase
      reaction rate, was determined for both normal and
      liver cirrhotic individuals in 30 mM veronal buffer
      pH 7.5 (section 10 of Experimental). The effect
      of various concentrations of nickel chloride was
      investigated at two substrate concentrations, linear
      relationship was frequently obtained as shown in
      Figures 10, 11. It was found that nickel ions act
      as competitive inhibitor to A – 5 – MP, the inhibitor
      constants were found to be markedly different with
      respect to normal and liver cirrhotic individuals
      (Table – 7).

      The degree of inhibition (defined as the percentage
      decrease in maximum velocity at each inhibitor
      concentration) was determined for both normal and

liver cirrhotic $\bar{5}$ - nucleotidase at 0.8 nM A - 5 - MP, it was found that there was a difference in the degree of inhibition in both cases (Figure - 12).

B -   Inhibition of $\bar{5}$ - nucleotidase by ATP, the reaction rate of $\bar{5}$ - nucleotidase was measured for both normal and liver cirrhotic individuals at the following concentrations of ATP, 0.1 mM, 0.2 nM, 0.4 nM, 0.6 nM, 0.8 mM, by using two substrate concentrations as described in section " 11 " of the " Experimental ".

The inhibition of $\bar{5}$ - nucleotidase by ATP was found to be of the competitive type, for both normal and liver cirrhotic persons. The inhibitor constants were found to be markedly different with respect to normal and liver cirrhotic sera (Table 7). Kinetic data were plotted using Dixon plot (Figures 13, 14). The degree of inhibition was determined at pH 7.5, and it was found to be much higher in normal than liver cirrhotic $\bar{5}$ - nucleotidase as shown in figure 15. The inhibition by ATP is dependent on its concentration for both systems as its shown in Figure 16.

C ..   Inhibition of serum $\bar{5}$ - nucleotidase by adenosine the velocity of $\bar{5}$ - nucleotidase was measured in

the presence of adenosine, at three different substrate concentrations. The range of concentrations
used were; 0.1 mM, 0.2 mM, 0.4 mM, 0.6 mM, 0.8 mM,
as explained in section 12 in the " Experimental ".

The inhibition of 5 - nucleotidase by adenosine was
found of the non - competitive type for both normal
and liver cirrhotic sera, (Figures 17, 18), the
inhibitor constants were found to be different with
respect to normal and pathological sera (Table 7).
The degree of inhibition for normal 5 - nucleotidase
at H 7.5, was much greater than for liver cirrhotic
sera as represented in Figure 19. The inhibition
by adenosine is dependent on its concentration for
both systems (Figure 20).

X -   Activation studies on 5 - nucleotidase

A -   Activation by Mg^{+2} for both normal and liver cirrhotic
       sera.

The role of 20 mM Mg^{+2} was studied and found to
activate 5 - nucleotidase of both normal and pathological sera. The Km values which were determined
by plotting v versus (S), found to be nearly equal
to the Km determined in the presence of 1.0 mM Mn^{+2}

83

(Figures 21, 22). The degree of activation (defined
as the percentage increase in maximum velocity at
each substrate concentration) was higher for pathological sera than normal $\bar{5}$ - nucleotidase at pH 7.5,
(Table - 8) a control experiment was performed
in the absence of any activating metal ions, and
the Km was found to be higher for both systems.

B   Activation by Mn^{+2} for both normal and liver cirrhotic
sera.

1.0 mM Mn^{+2} was found to activate $\bar{5}$ - nucleotidase
of both systems, the Km values which were determined
in the presence of 1.0 mM Mn^{+2} were higher for normal
than liver cirrhotic sera. A control experiment was
performed, from which the Km was determined in the
absence of Mn^{+2}, and it was found to be higher for
both normal and cirrhotic individuals.

X -- The effect of temperature on $\bar{5}$ - nucleotidase activity
in normal and liver cirrhotic sera.

Different incubation temperatures were used for 30
minutes at optimum conditions of substrate concentration
(.0 mM A - 5 - MP) and at pH 7.5. The temperatures
employed were; 8^{0}, 25^{0}, 37^{0}, 45^{0}, 60^{0}, 100^{0}, the results
obtained are represented in Figures 25 , 26. Figure 25

shows the effect of temperature on the rate of the ezymatic
reaction and figure 26 is the Arrhenious plot of the same
data, The experiment was run as described in the " Experimental ", and the activity was measured as I.U./liter.

XII - pH studies on 5 - nucleotidase in normal and cirrhotic
        sera.

Tho pH effect on this enzyme was studied in both
normal and cirrhotic sera at 37°, using veronal buffer.

A -   Relationship of pH and the velocity of the reaction
        at different A - 5 - MP concentrations.

        The velocity of the reaction was measured at different
        A - 5 - MP concentrations in the pH range (7 - 8.5);
        it was found that, there was a pH optimum at each
        A - 5 - MP concentration used when A - 5 - MP con-
        centrations were varied between 0.2 - 0.8 mM, these
        pH optimum shifted towards higher pH vaules with
        increasing A - 5 - MP concentration, these values
        and other informations are presented in Figure 27.

B -   Relationship of pH and the optimum A - 5 - MP concen-
        trations.

        It was observed that there was an optimum A - 5 - MP
        concentration at each pH value; when the velocity

of $\bar{5}$ - nucleotidase was measured at different pH
values in presence of different A - 5 - MP concentrations, the optimum A - 5 - MP concentration
shifted towards higher values, while the pH was
increased from 7 - 8.5 in normal individuals, this
is not the case when serum of liver cirrhotic
patients was used, as represented in Figure 28.

C -   The effect of pH on the initial velocity.

Initial velocity was determined at different pH
values in presence of two different A - 5 - MP
concentrations for both normal and liver cirrhotic
sera. Results which are presented in Figure 29,
show the relationship between log v versus pH, in
the pH range (7 - 8.5).

D -   The effect of pH on the Km value.

Km (A - 5 - MP) for $\bar{5}$ - nucleotidase was determined at various pH values in the veronal buffer
Km (A - 5 - MP) was found to be very much pH
dependent at pH values higher than 7; (Figure 30),
pK_e for $\bar{5}$ - nucleotidase was determined.

E —   pH optimum determination in the presence and absence
of 20 mM Mg^{+2} ions; at 37°.

The reaction rate was determined at 1.0 mM A — 5 — MP
and varying pH (7 — 8.5) using veronal buffer, in
the presence and absence of 20 mM Mg^{+2}, for both
normal and liver cirrhotic sera. The results which
are represented in Figures 31, 32, show that, the
velocities at the respected pH optimum were higher
in the presence of Mg^{+2} ions than in its absence,
for both systems.

XIII —   Studies on the nature of $\bar{5}$ — nucleotidase in both
normal and liver cirrhotic sera at 37°.

A —   Relationship between substrate concentration and
the velocity of the reaction in both normal and
cirrhotic sera.

Figure 4 shows the hyperbolic relationship between
the velocity and substrate concentration, so the
curve is obeying Michaelis — Menten equation, in
the absence of any inhibitor of the nucleotides
nature.

3 - Determination of interaction coefficient n, between
inhibitor binding sites, for both normal and liver
cirrhotic sera.

If Hill[141] plot of coordinate is applied to kinetic
measurements of $\bar{5}$ - nucleotidase by plotting
$\log \dfrac{v}{V-v}$ against $\log$ (A - 5 - MP) concentration, a straight line of positive slope $n = 1$ was
obtained (see Figure 33), when the Hill system
of coordinate is applied to kinetic measurements
of $\bar{5}$ - nucleotidase, by plotting $\log \dfrac{v_i - v_i \text{ (Sat)}}{v_o - v_i}$
(where v_i is the reaction velocity in the presence
of the inhibitors, v_i (Sat) is the reaction velocity
at the saturating concentration of inhibitors, and
v_o is the reaction velocity in the absence of inhibitors) against log of inhibitors concentrations (ATP),
a straight line of negative slope n, higher than
one was obtained for normal sera, but $n = 1$ for
liver cirrhotic sera, as represented in Figure 34.

T A B L E S

Table - 1. $\bar{5}$ - nucleotidase activity in normal and liver cirrhotic sera at 37°.

The method for the assay of $\bar{5}$ - nucleotidase was described in section A of the experimental, analysis were compensed on the same day of sample collection. The activity measurements were expressed in I.U./liter.

Specimen	No.of cases	Age group years	Activity range I.U./L.	Average activity I.U./L.	% Increase in activity over normal
Normal	60	20-40	1.45-12.09	7.80	0.00
Liver cirrhotic *	40	12-60	5 - 50	18.30	57.37
Biliary cirrhosis	8	12-60	20-50	32.77	76.19

* Untreated

Table - 2. The effect of sex on the activity of
5 - nucleotidase, parallel to the age groups.

The activity of 5 - nucleotidase was measured
at 37°, using optimum reaction conditions, as
described under section A of the " Experimental ".

Specimen	Age (years)	Sex	Activity range I.U./liter
Normal	20 - 30	Female	1.45 - 12.09
	20 - 30	Male	2.04 - 12.09
	30 - 40	Female	4.01 - 11.60
	30 - 40	Male	7.00 - 12.00
Liver cirrhosis	10 - 20	Female	4.90 - 13.15
	10 - 20	Male	17.00 - 18.70
	20 - 30	Female	19.04 - 19.50
	20 - 30	Male	13.15 - 19.90
	30 - 60	Female	17.00 - 19.04
	30 - 60	Male	4.11 - 19.90

Table - 3. Stability of 5 - nucleotidase for both
normal persons and patients with liver
cirrhosis.

Sera from both normal and liver cirrhotic
patients, were kept for 30 min. at the
following temperatures, 27°, 45°, 49°, 52°,
60°. Then the activity was measured as
described under " Experimental ", Section A,
for the enzyme assay.

Temperature	Normal activity I.U./L.	% Decrease in activity	Liver cirrhotic activity I.U./L.	% Decrease in activity
27°	7.89	0.00	9.96	0.00
45°	5.40	31.55	2.50	74.89
49°	2.91	63.11	0.42	95.78
52°	1.66	78.96	0.23	97.69
60°	0.05	99.36	0.01	99.89

Table - 4. Metal requirements

> The activity of 5 - nucleotidase was measured
> in both normal and liver cirrhotic sera, in
> the presence of different metal ions, as 20mM
> Mg^{+2}, 1 mM Mn^{+2}, 10 mM Ni^{+2}. The method
> used was as described in " Experimental ",
> the results were expressed in I.U./Liter.

Specimen	Ions added			
	None	Mg^{+2}	Mn^{+2}	Ni^{+2}
Normal	3.30	4.30	6.00	0
Liver cirrhosis	14.01	20.00	22.45	0

Table - 5.   Determination of K_m (A - 5 - MP) for

$\overline{5}$ - nucleotidase in normal and liver cirrhotic

sera.

The reaction was carried out in 30 mM veronal
buffer pH 7.5 at 37°. The velocity of the
reaction was determined at the following substrates concentrations, 0.05 mM, 0.1 mM, 0.2 mM,
0.4 mM, 0.6 mM, 0.8 mM, 1.0 mM .

The experimental data (v and (S)) were plotted
in various forms, the new direct linear plot was
compared with the traditional methods currently
in use.   Values presented in the following table
are of 6 samples + standard deviation.

	K_m (A - 5 - MP) nM			
Specimen	Method of plotting			
	(a) v vs. (S)	(b) $\frac{1}{v}$ vs. $\frac{1}{(S)}$	(c) $\frac{(S)}{v}$ vs. (S)	(d) Direct linear plot
Normal serum	0.195 + 0.024	0.2 + 0.01	0.21 + 0.02	0.18 + 0.03
Liver cirrhotic serum	0.0533 + 0.02	0.07 + 0.01	0.08 + 0.01	0.06 + 0.02

(a)   The plot is according to the original Michaelis - Menten
equation $v = \dfrac{V (S)}{K_m + (S)}$

(b)   The Lineweaver - Burk double reciprocal plot
$\dfrac{1}{v} = \dfrac{K_m}{V} \cdot \dfrac{1}{(S)} + \dfrac{1}{V}$

(c)   The method of Hans $\dfrac{(S)}{v} = \dfrac{K_m}{V} + \dfrac{1}{V} (S)$

(d)   Refer to " Experimental " for details

Table - 6. <u>Activity of 5 - nucleotidase in normal and liver cirrhotic persons, measured in the presence of 2 - d - A - 5 - MP.</u>

Activity of 5 - nucleotidase was calculated in the presence of different concentrations of 2 - d - A - 5 - MP, 0.1 mM, 0.2 mM, 0.4 mM, 0.6 mM, 0.8 mM, 1.0 mM, and 30 mM veronal buffer pH 7.5, at 37°. As described in section 4 of " Experimental ".

A -

Substrate concentration mM	Activity for liver cirrhotic I.U./L.	Activity for normal I.U./L.	Ratio of liver cirrhotic to normal
0.2	5.23	1.89	2.76
0.4	7.20	2.23	3.22
0.6	10.95	3.51	3.11
0.8	11.65	4.70	2.47

B -

Specimen	Km (mM)
Normal	0.35
Liver cirrhotic	0.25

Table – 7. <u>Inhibition constants for 5 – nucleotidase in both normal and liver cirrhotic individuals.</u>

The rate of the reaction was determined in the presence of varying concentrations of the specified inhibitor at two substrate concentrations as described before the type of inhibition and the inhibitor constants were determined from plots of $\frac{1}{v}$ versus (i). The values presented in the table are of 3 samples + standard deviation.

Specimen	Inhibition constant (Ki) for :-		
	Nickel	ATP	Adenosine
Normal	0.26 + 0.031	0.1 + 0.03	0.47 + 0.09
Liver cirrhotic	0.315 + 0.012	0.225 + 0.02	0.59 + 0.08

Table ~ 8. The degree of activation of 5 ~ nucleotidase by Mg^{+2} ions, at 37°.

The degree of activation[*] of 5 ~ nucleotidase was determined in the presence of 30 mM veronal buffer pH 7.5, and at 0.8 mM A ~ 5 ~ MP, the concentration of Mg^{+2} ions employed was 20 mM. The activity of 5 ~ nucleotidase was expressed in I.U./liter for both normal and liver cirrhotic sera. The results are the average of 3 samples

Specimen	Degree of activation
Normal serum	12.5%
Liver cirrhotic serum	28.5%

[*] The degree of activation =

$$\frac{\text{Velocity in presence of } Mg^{+2} \text{ ions } - \text{Velocity in absence of } Mg^{+2}}{\text{Velocity in presence of } Mg^{+2} \text{ ions}}$$

x 100

FIGURES

96

Figure - 1. Relationship of $\overline{5}$ - nucleotidase activity
with time, in normal individuals, at 37°

The velocity of $\overline{5}$ - nucleotidase was measured
at different times intervals, ranging between
5 to 35 min., at 1.0 mM A - 5 - MP, 30 mM
veronal buffer pH 7.5. The activity was
measured according to the method described
in point number 2, of the " Experimental ".

Fig.1

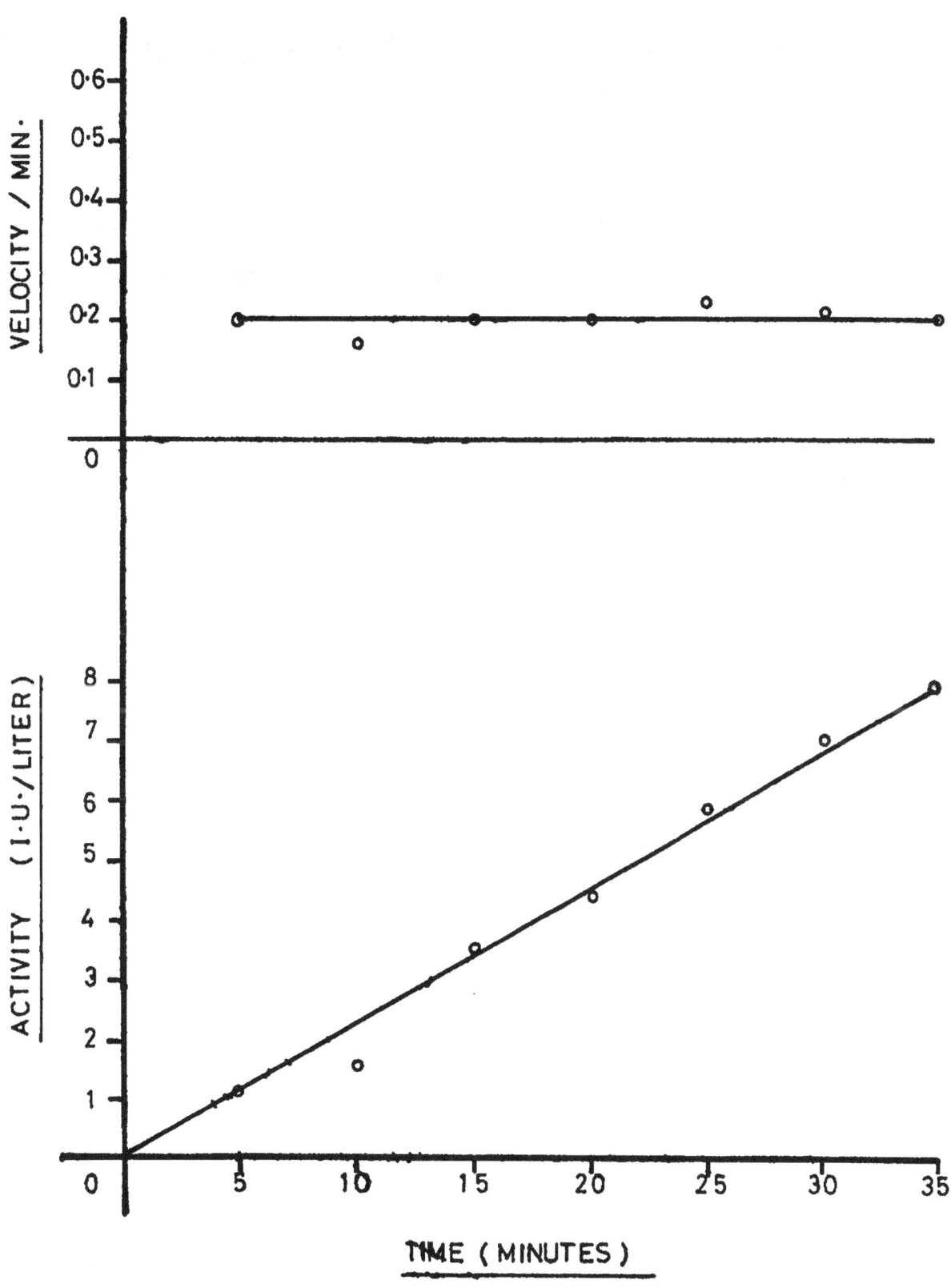

Figure - 2.   Relationship of $\bar{5}$ - nucleotidase activity
with time intervals at 37°, for liver
cirrhotic individuals.

The velocity of $\bar{5}$ - nucleotidase was measured
at different time intervals, ranging between
5 to 35 min., details are as under " Experimental "

Fig·2

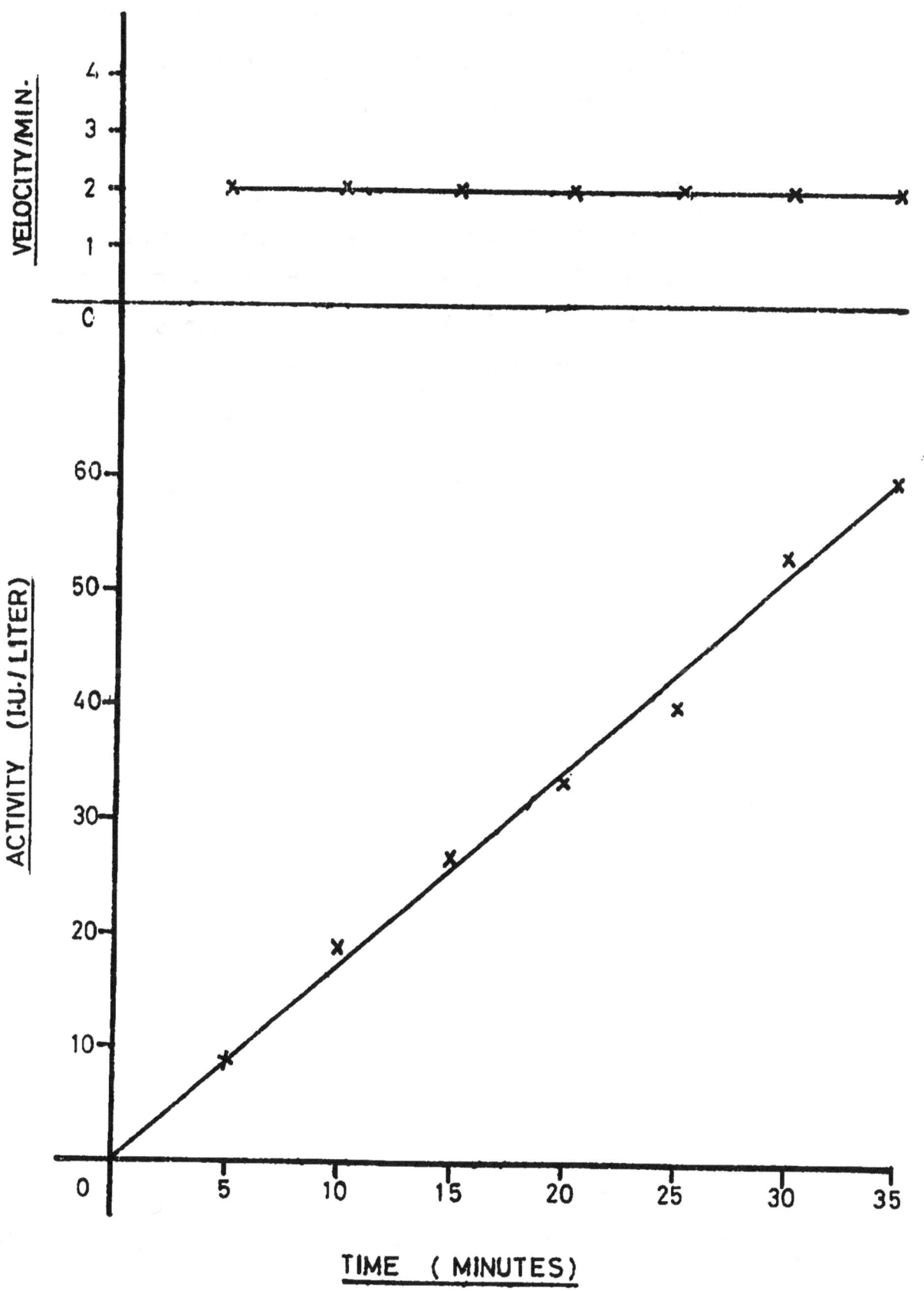

TIME (MINUTES)

· 100

Figure - 3.   Optimal A - 5 - MP concentration for
5 - nucleotidase activity for both normal
and liver cirrhotic sera

The reaction was carried out at 37° in the
presence of 30 mM veronal buffer pH 7.5,
and different concentrations of A - 5 - MP,
ranging from 0.05 mM to 1.0 mM, the 5 - nucleotidase activity was calculated as I.U./liter
Details are as under " Experimental ".

x ———x, normal;▲ ——————▲ liver cirrhotic
serum.

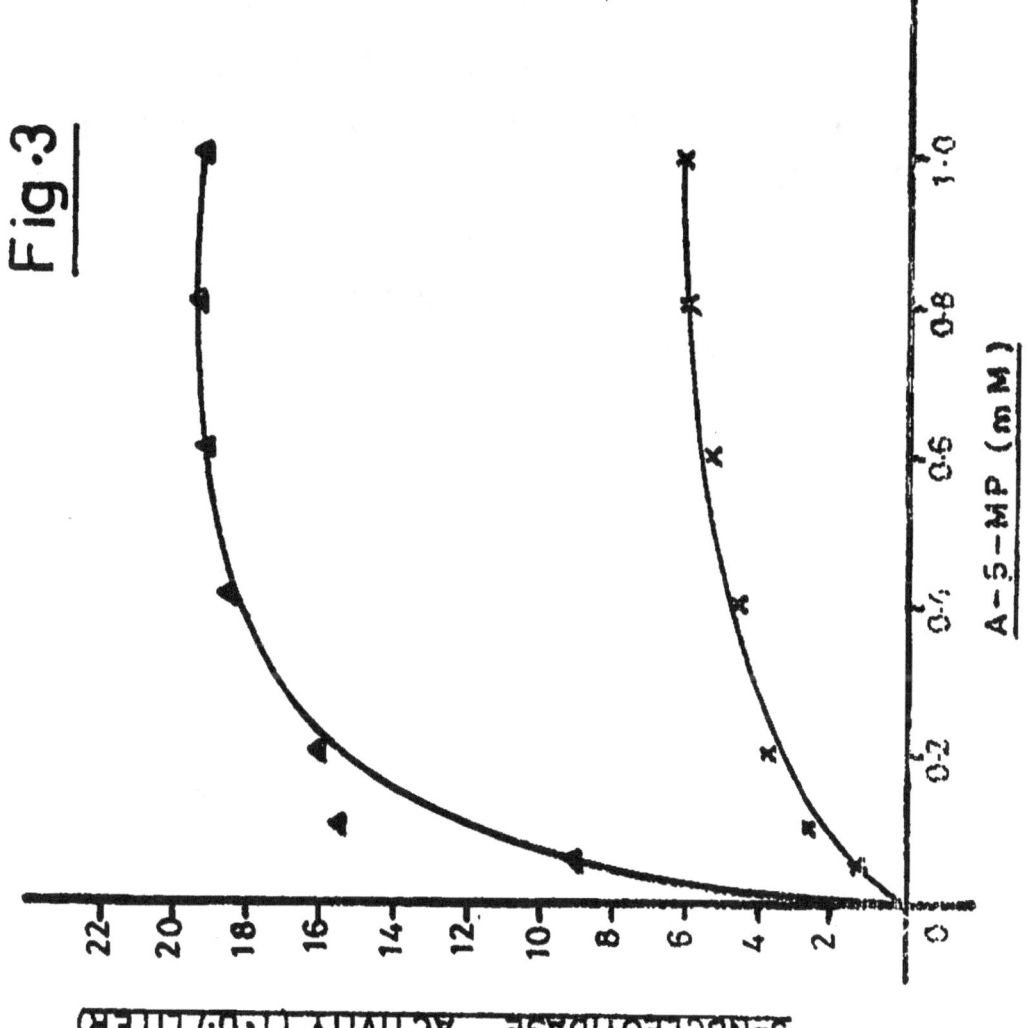

Fig.3

A-5-MP (mM)

5-NUCLEOTIDASE ACTIVITY (I.U./LITER)

Figure - 4.   Km (A - 5 - MP) determination for 5 - nucleotidase by plotting v versus (S)

The rate of the reaction was measured at
37°, and 30 mM veronal buffer pH 7.5, using
different substrate concentrations, 0.05 mM,
0.1 mM, 0.2 mM, 0.4 mM, 0.6 mM, 0.8 mM, 1.0 mM.
The velocity of the reaction was calculated
as I.U./liter. The equation for this kind
of plotting is

$$v = \frac{V\,(S)}{Km + (S)}$$

Details are as under " Experimental ".

x ———— x, normal serum; ●——————● liver
cirrhotic serum.

Fig.4

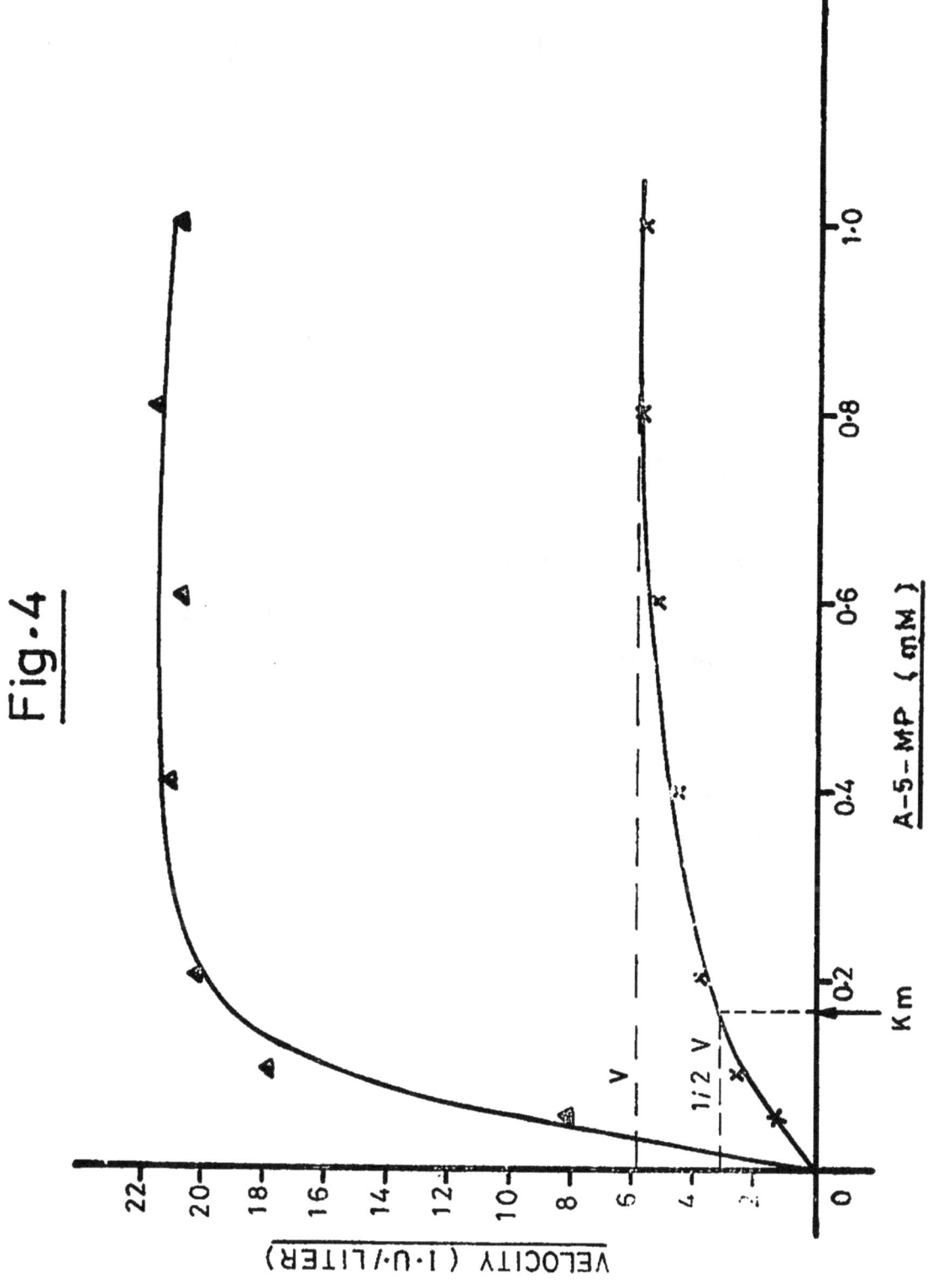

Figure - 5.   Km (A - 5 - MP) determination for serum
5 - nucleotidase by Hans method, which is
a modification of Michaelis - Menten equation; $\frac{(S)}{v} = \frac{Km}{V} + \frac{1}{V}$ (S).

$\frac{(S)}{v}$ is plotted versus (S), reaction conditions are, 30 mM veronal buffer, pH 7.5,
A - 5 - MP concentrations ranging between
0.05 mM - 1.0 mM, at 37°.   Details are as
described in section 3.

. ———— ., for normal serum, x————x for
liver cirrhotic serum.

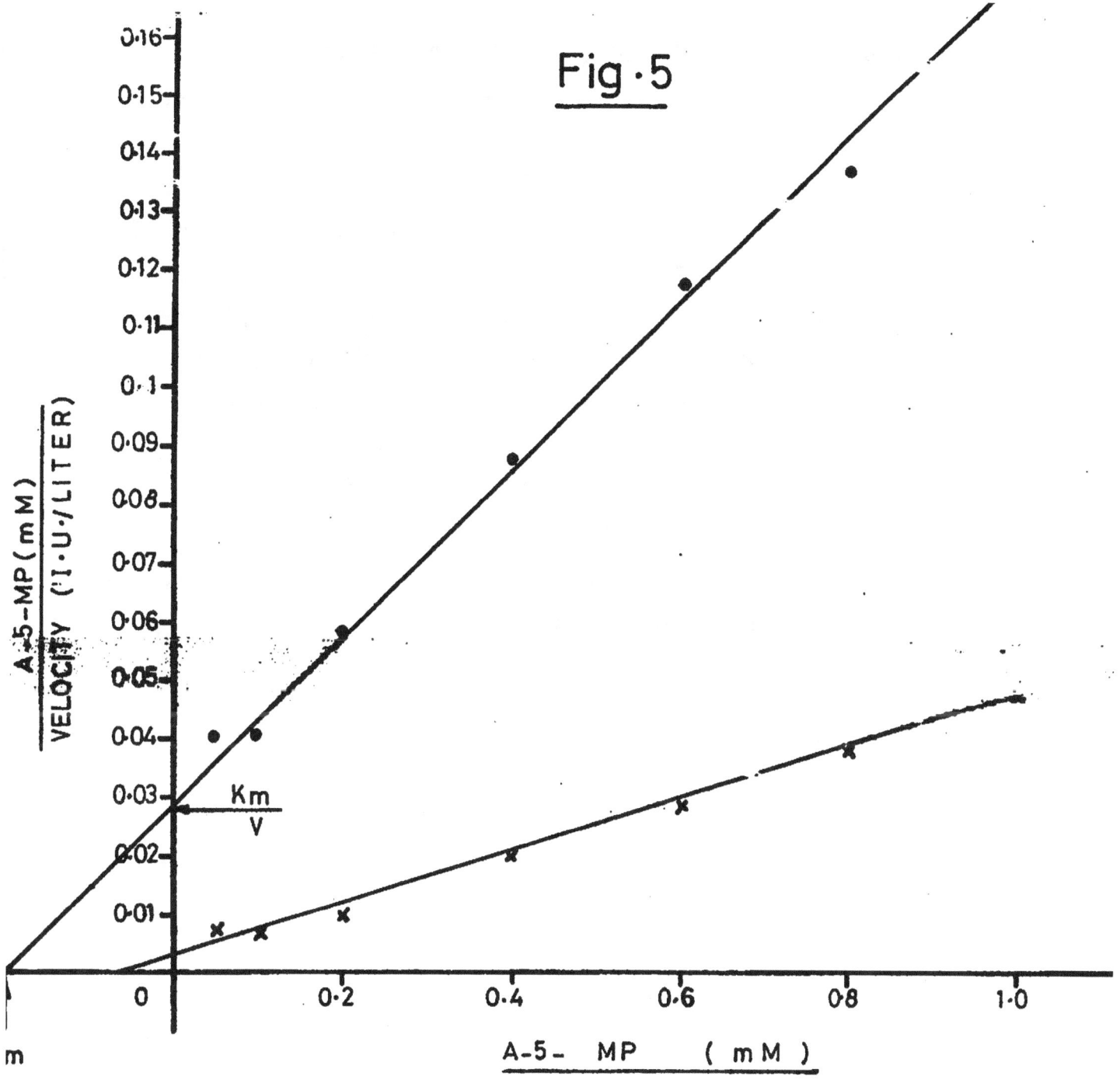

Fig ·5

Figure – 6.   Km (A – 5 – MP) determinations for serum
             5 – nucleotidase using direct linear plot.

This method of plotting was explained under
" Experimental ".   The reaction was carried
out at 37°, 30 mM veronal buffer pH 7.5.
The velocity was calculated in I.U./liter,
at the following A – 5 – MP concentrations;
0.2 mM, 0.4 mM, 0.6 mM, 0.8 mM, 1.0 mM.

A – is a graph for determining Km (A – 5 – MP)
in liver cirrhotic serum.

B – is a graph for determining Km (A – 5 – MP)
in normal serum.

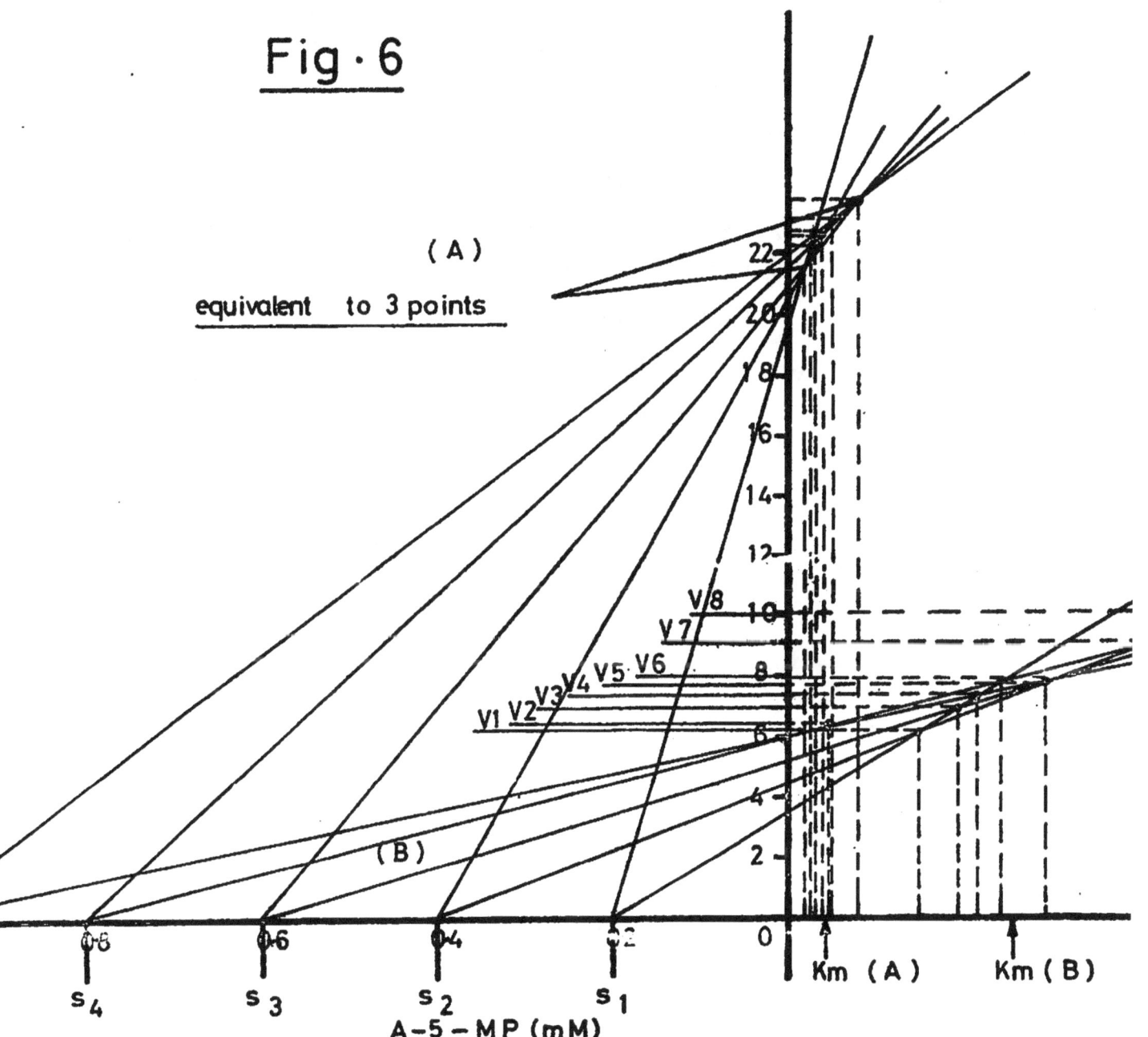

Fig · 6

(A)

equivalent to 3 points

Figure - 7.   Km (A - 5 - MP) determination for 5 - nucleo-
              tidase by using Lineweaver double reciprocal
              plot.

The velocity of the reaction was determined
at 37°, using 30 mM veronal buffer pH 7.5,
and ranging substrate concentrations, from
0.05 mM to 1.0 mM A - 5 - MP, as described
in section 3 of the experimental.   The
equation used for plotting $\frac{1}{v}$ versus $\frac{1}{(S)}$
is

$$\frac{1}{v} = \frac{Km}{V} \cdot \frac{1}{(S)} + \frac{1}{V}$$

▲ — — — ▲ , normal serum, X——————X liver
cirrhotic serum.

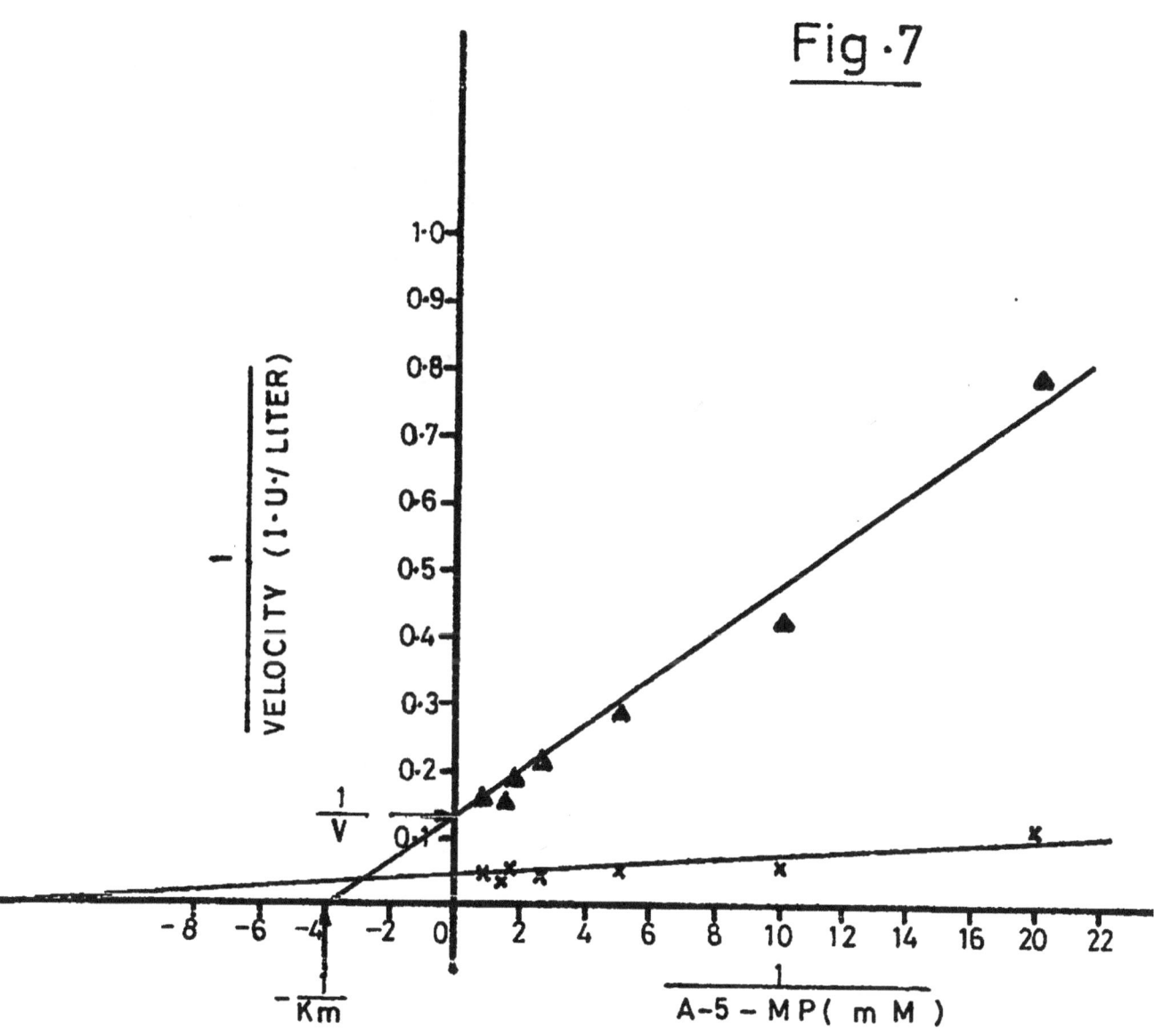

Fig ·7

Figure - 8. Km ($\bar{2}$ - d - A - 5 - MP) determination for normal and liver cirrhotic patients, at 37°.

Km ($\bar{2}$ - d - A - 5 - MP) was determined in the presence of varying concentration of $\bar{2}$ - d - A - 5 - MP ranging between 0.1 - 1.0 mM, at 30 mM veronal buffer pH 7.5, as illustrated under " Experimental ".

.———— .; for normal; o ———— o for liver cirrhotic sera.

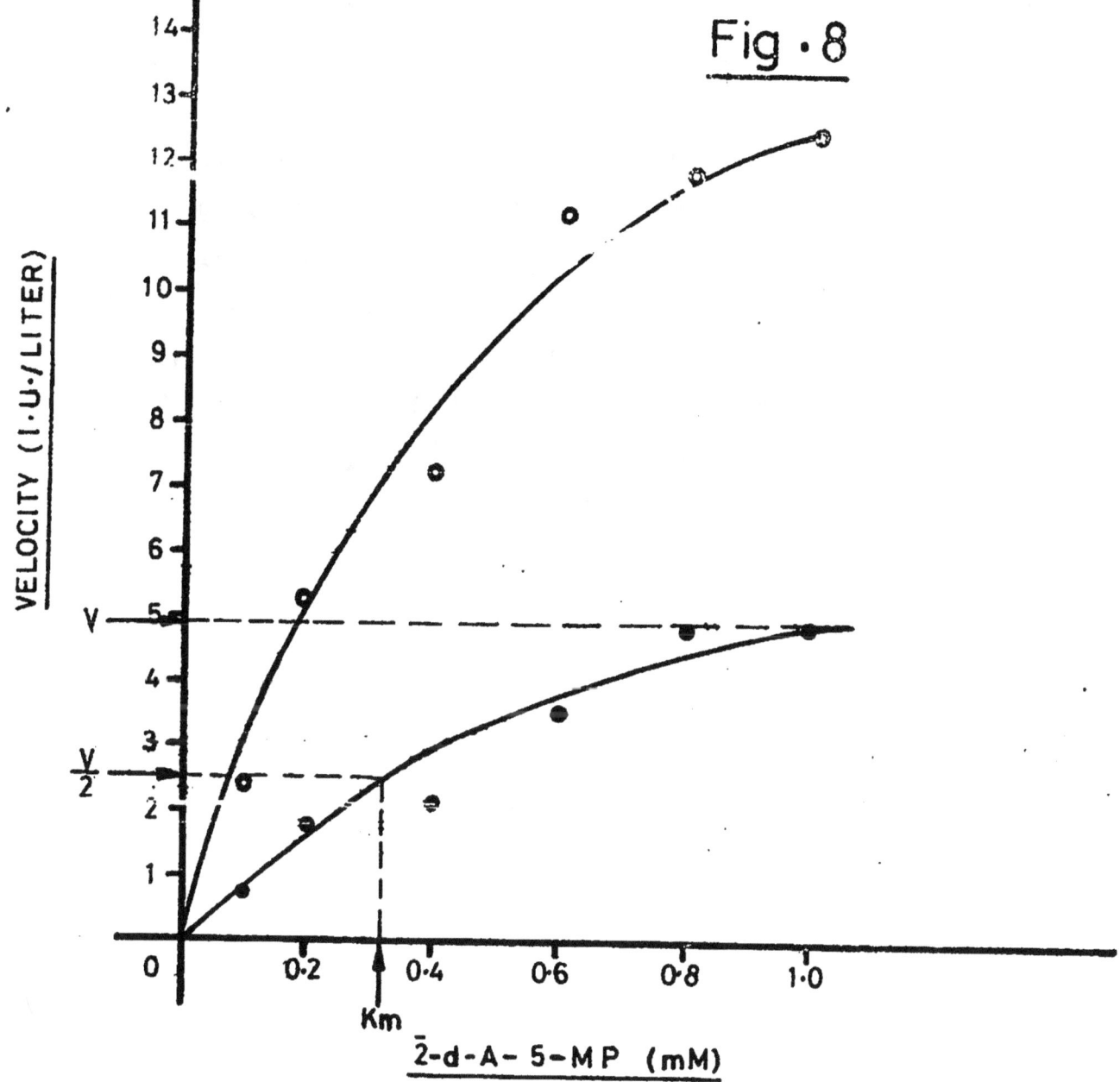

Fig · 8

112

Figure – 9. Variation of serum concentrations with
5 – nucleotidase activity .

Activity was calculated as I.U./liter for
both normal and liver cirrhotic sera, the
conditions for measurements were optima.
Different concentrations of serum were
employed ranging between 50 /uL – 300 /uL.
Details are as written under " Experimental "

x ——————— x; normal serum; o ——————— o ; for
liver cirrhotic serum.

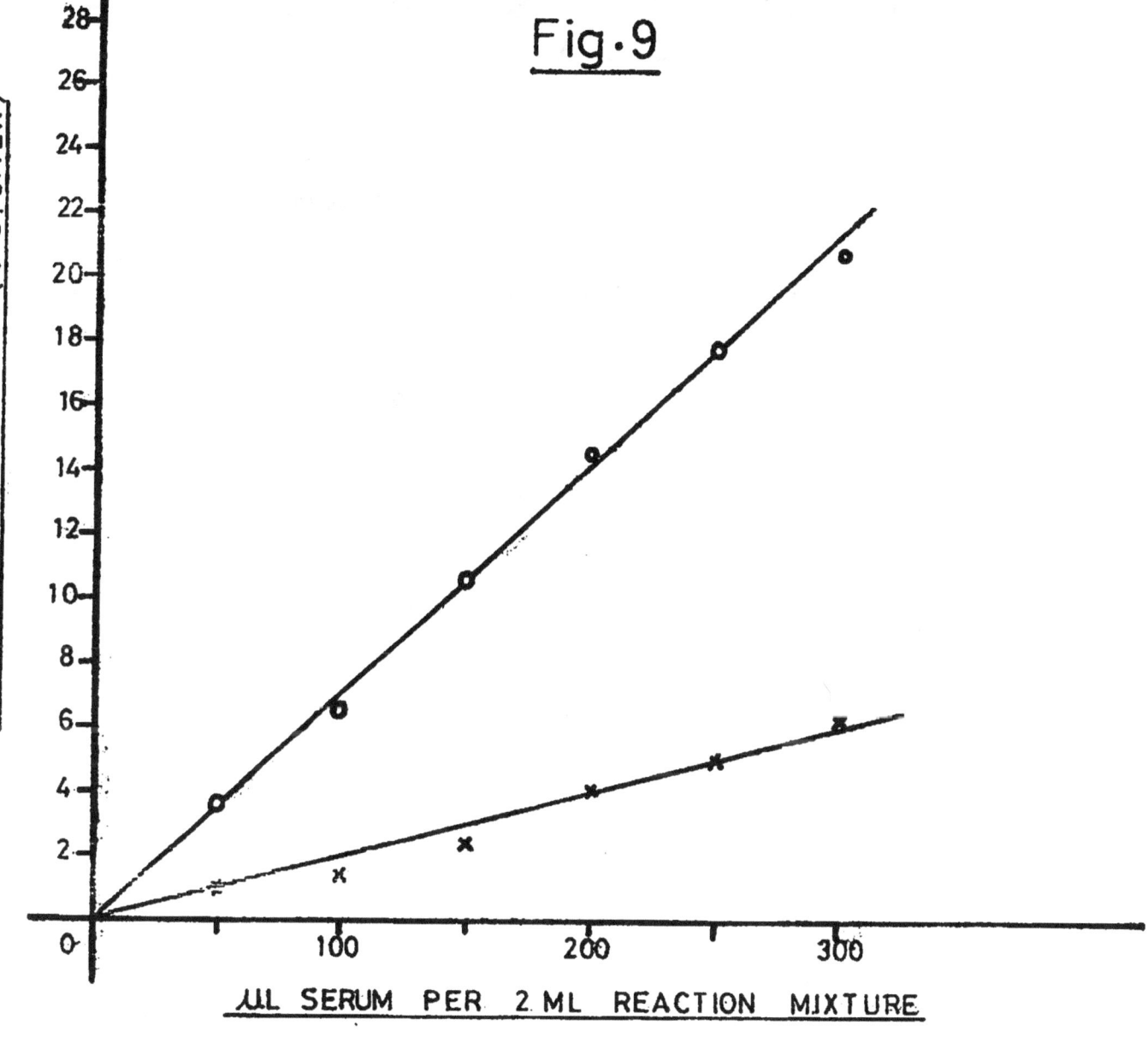

Fig.9

μL SERUM PER 2 ML REACTION MIXTURE

Figure 10.   Inhibition of 5 - nucleotidase by Nickel
              ions in normal persons.

              The velocity of the reaction was measured at
              37°, over a range of Ni^{+2} concentrations from
              0.1 - 0.4 mM, at 0.4 and 0.8 mM A - 5 - MP.
              Dixon method of plotting was used.

              o ———— o, for 0.8 mM A - 5 - MP, x ———— x
              for 0.4 mM A - 5 - MP.

Fig·10

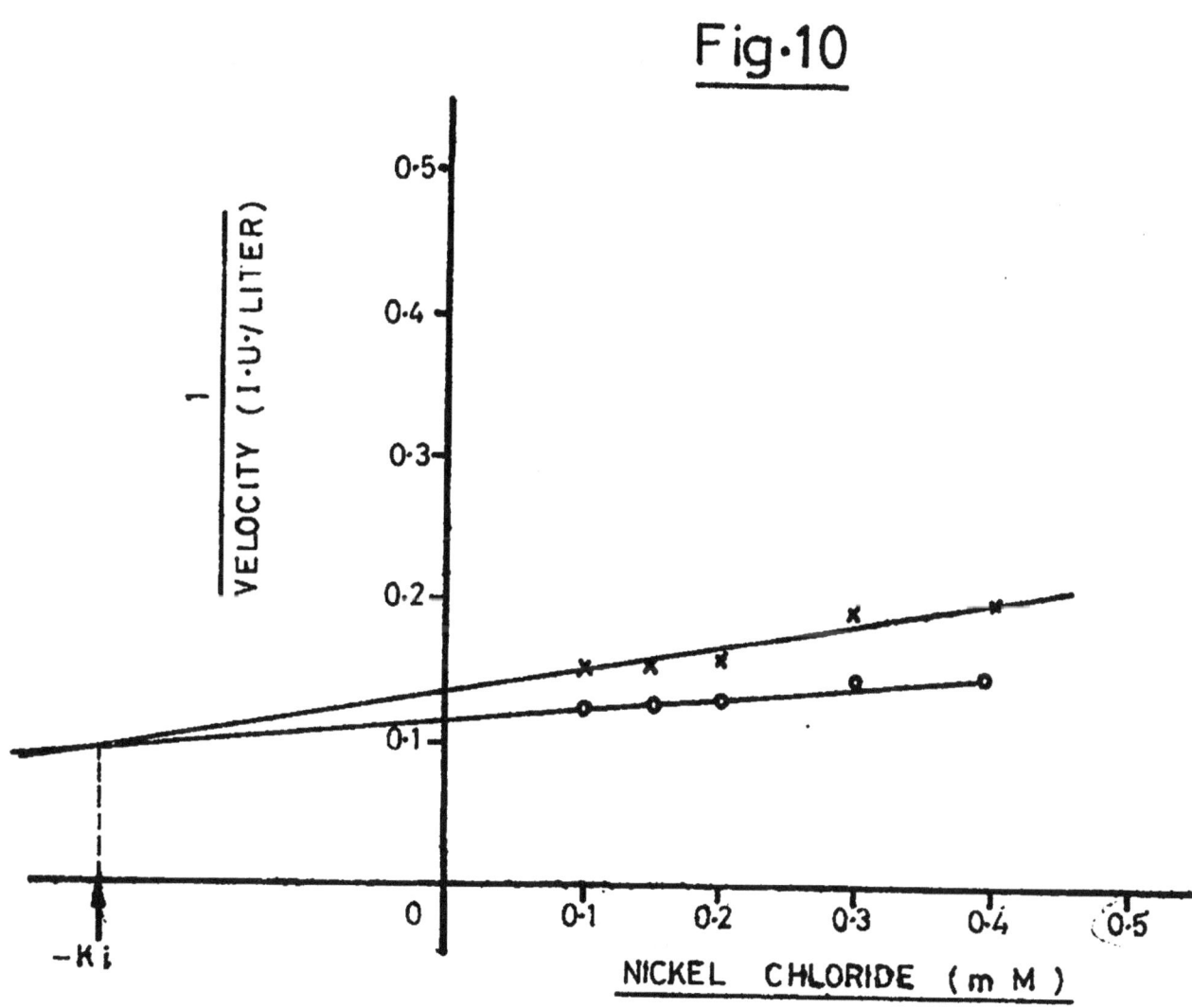

NICKEL   CHLORIDE   (m M)

Figure - 11.   Inhibition of 5 - nucleotidase of liver
cirrhotic serum by Nickel ions.

Details are as in " Figure 10 ".
Reaction rate was determined at pH 7.5.

o ------- o, for 0.8 mM A - 5 - MP, x ------- x
for 0.4 mM A - 5 - MP.

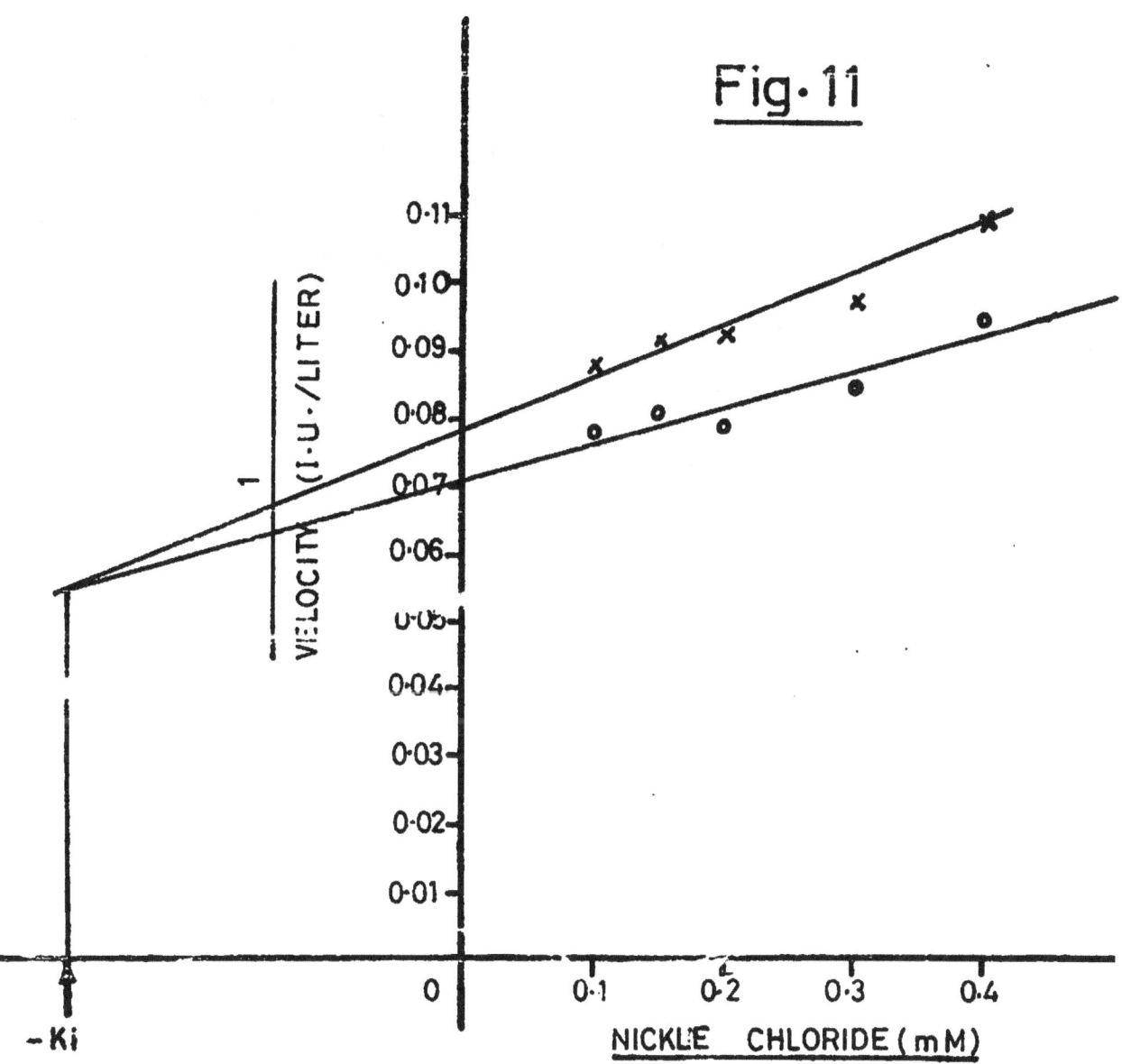

Fig· 11

Figure - 12.   The degree of inhibition for Ni^{+2}, at 37^{0}

The rate of $\bar{5}$ - nucleotidase reaction was
determined in the presence of varying
concentrations of Ni^{+2}; 0.1 mM, 0.15 mM,
0.2 mM, 0.3 mM, 0.4 mM, at 0.8 mM A - 5 - MP,
pH 7.5, for both normal and liver cirrhotic
sera.

. ———— ., pathological, x ———— x for
normal.

Fig ·12

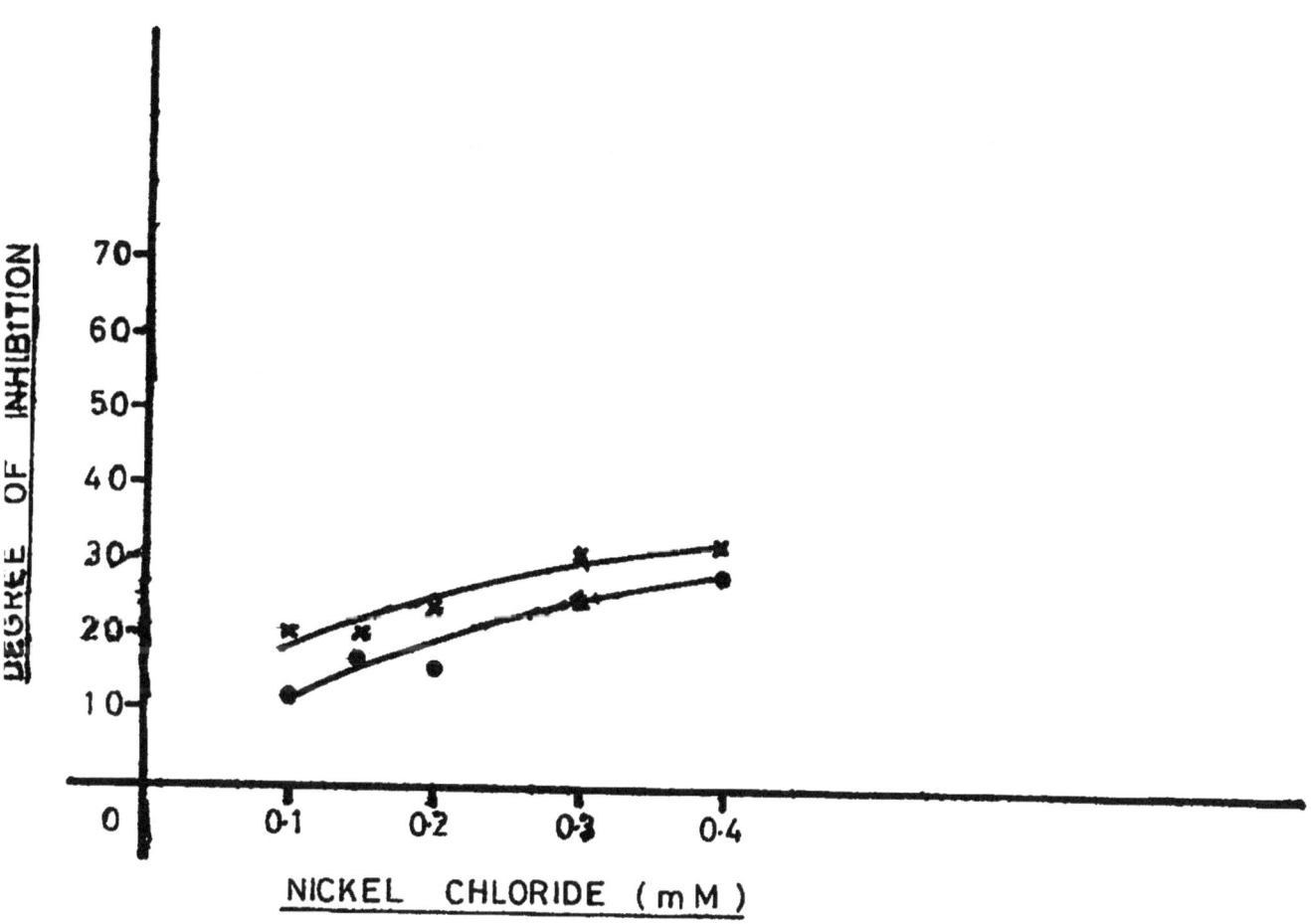

Figure – 13. The inhibition of 5 – nucleotidase by ATP for normal individuals.

The velocity of the reaction was measured at 37°, and in the presence of 0.1 mM, 0.2 mM, 0.4 mM, 0.6 mM, 0.8 mM, ATP, at two substrate concentrations (0.6 mM and 1 mM).
Dixon method of plotting was used.

x ——— x, for 1 mM A – 5 – MP, o ——— o, for 0.6 mM A – 5 – MP.

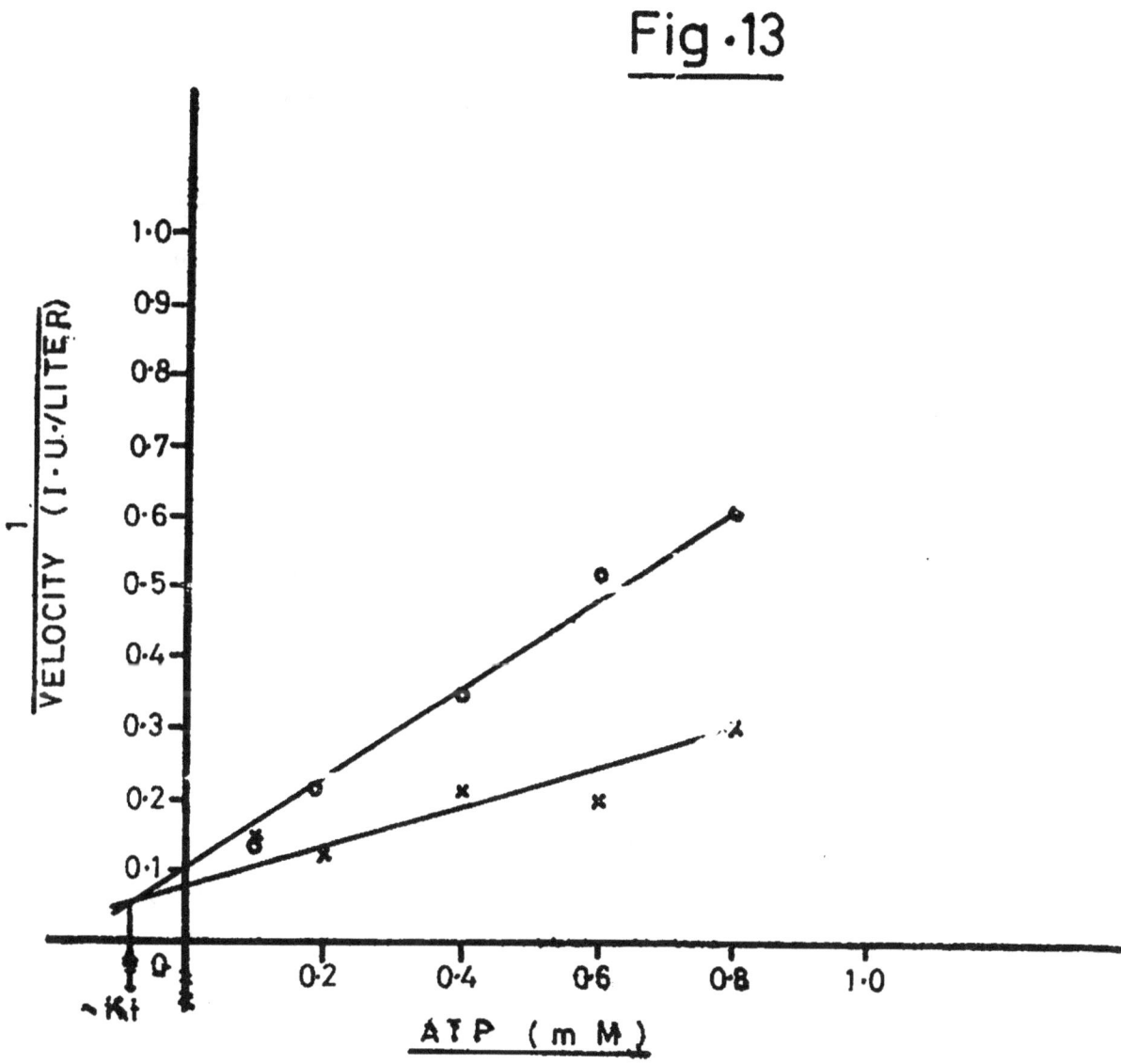

Fig ·13

Figure – 14. The inhibition of $\overline{5}$ – nucleotidase by ATP for liver cirrhotic individuals.

Details are as in " Figure – 13 ".

x ————— x, for 1.0 mM A – 5 – MP,

o ————— o, for 0.6 mM A – 5 – MP.

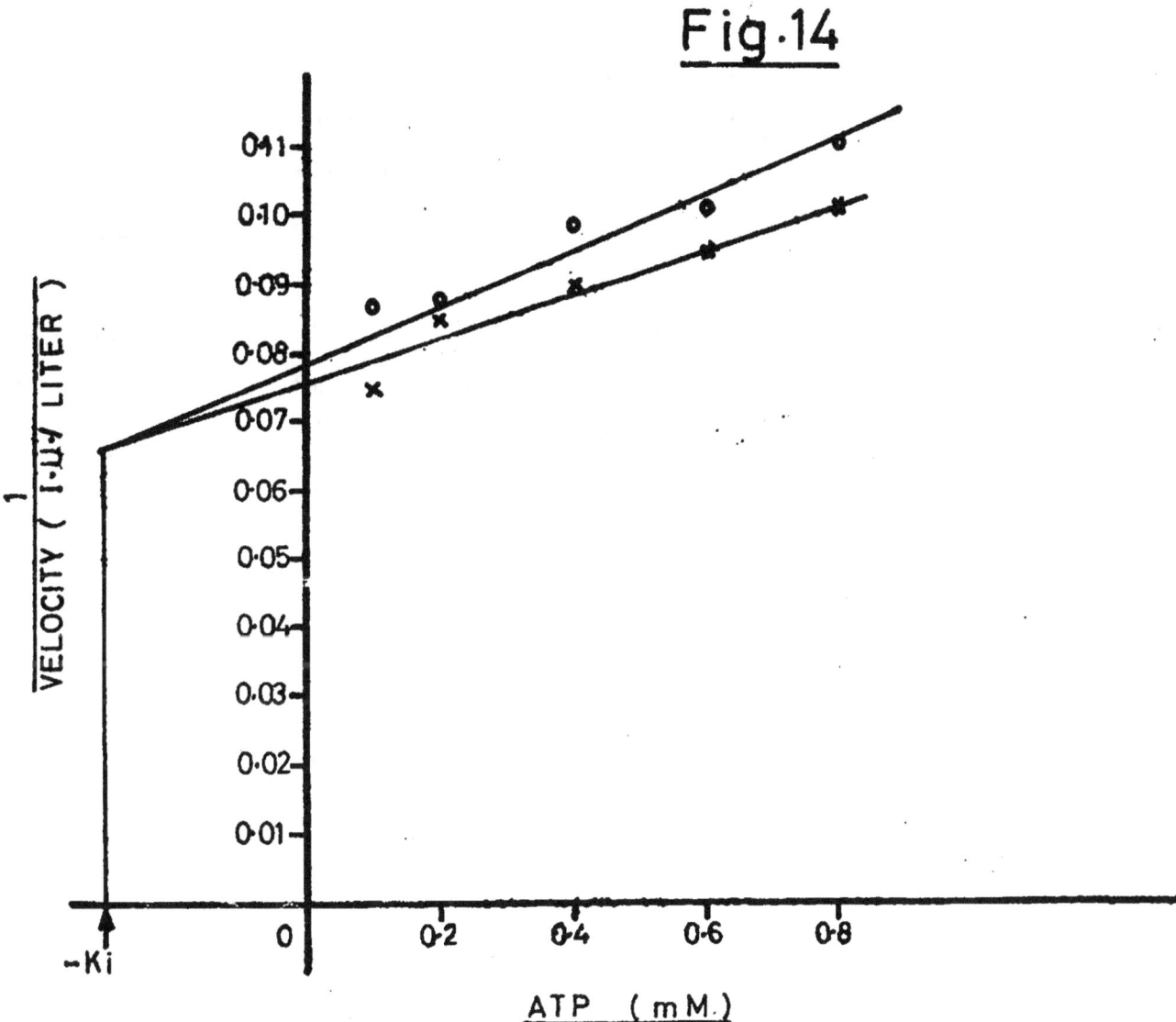

Fig.14

Figure - 15.   The degree of inhibition of $\bar{5}$ - nucleotidase
by ATP in normal and liver cirrhotic sera

The degree of inhibition[*] was determined
from velocity measurements in 30 mM veronal
buffer pH 7.5, at varying concentrations
of ATP, Details are as under " Experimental ".

△ ———— △ for 0.6 mM A - 5 - MP, ▲ ———— ▲
for 1.0 mM A - 5 - MP (in liver cirrhotic
sera), x ——— x for 0.6 mM A - 5 - MP,
o ——— o, for 1.0 mM A - 5 - MP (in
normal sera).

* Defined as the percentage decrease in
  maximum velocity at each ATP concentration.

Fig·15

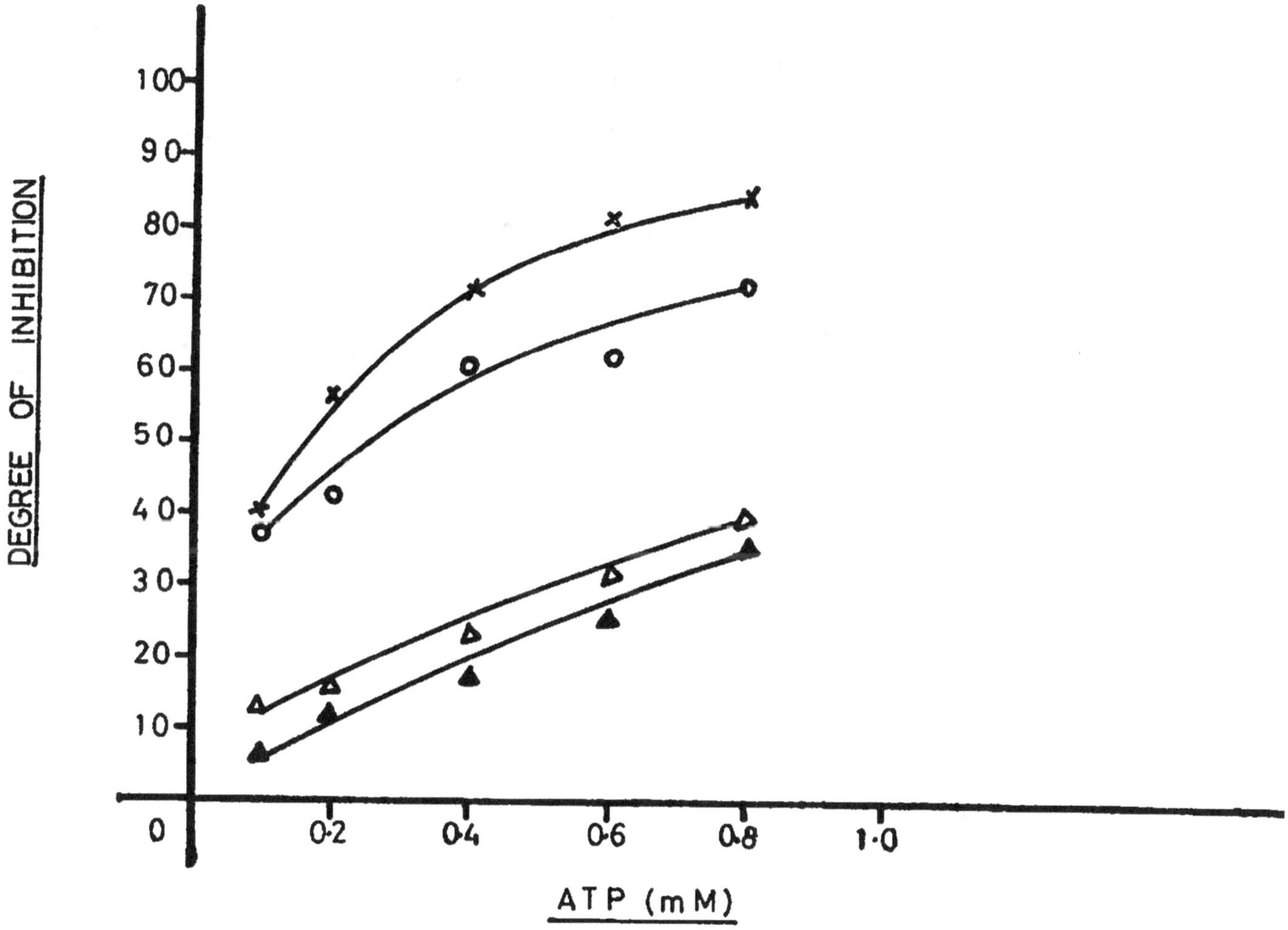

DEGREE OF INHIBITION

100
90
80
70
60
50
40
30
20
10

0   0·2   0·4   0·6   0·8   1·0

ATP (mM)

Figure - 16.   The velocity of the reaction was plotted
against different ATP concentrations for
both normal and liver cirrhotic sera, at 37°.

The velocity of the reaction was determined
for both normal and liver cirrhotic sera ,
in the presence of different concentration
of ATP from 0.1 -- 0.8 mM.   At pH 7.5.

o ———— o for 0.6 mM A - 5 - MP, x ——— x
for 1.0 mM A - 5 - MP (normal serum),
▲————▲ for 0.6 mM A - 5 - MP, .————.
for 1.0 mM A - 5 - MP (liver cirrhotic sera).

Fig.16

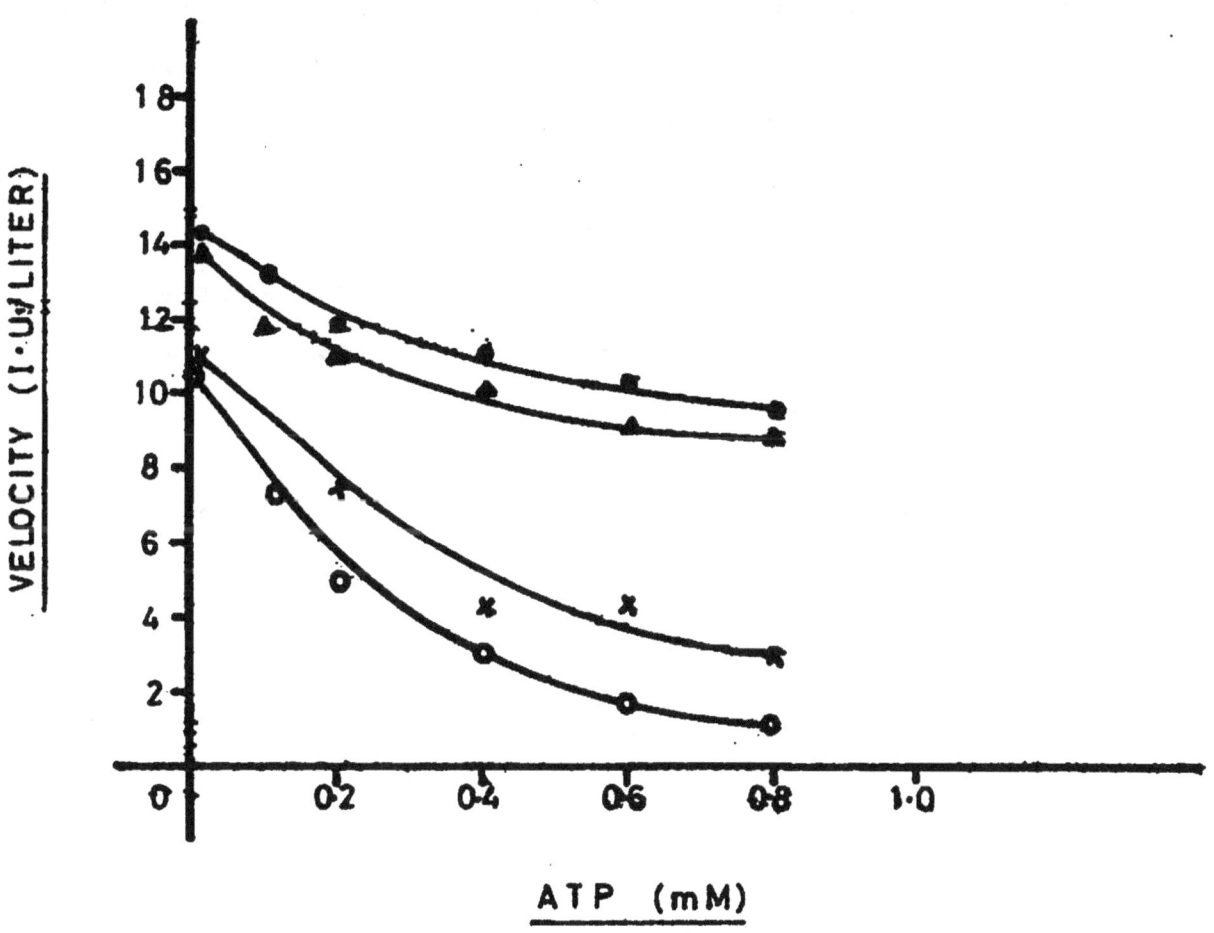

Figure - 17.   Inhibition of $\bar{5}$ - nucleotidase by adenosine for normal serum.

The effect of different concentrations of adenosine on the reaction rate of $\bar{5}$ - nucleotidase was studied, using varying concentrations of adenosine, 0.1 mM, 0.2 mM, 0.4 mM, 0.6 mM, 0.8 mM, at 0.6 mM and 1.0 mM A - 5 - MP, the pH used was 7.5.
Ki was determined by using Dixon plot, which is the plot of $\frac{1}{v}$ versus adenosine concentrations at two different substrate concentrations.

x ———— x, for 1.0 mM A - 5 - MP, o ——— o for 0.6 mM A - 5 - MP.

Fig·17

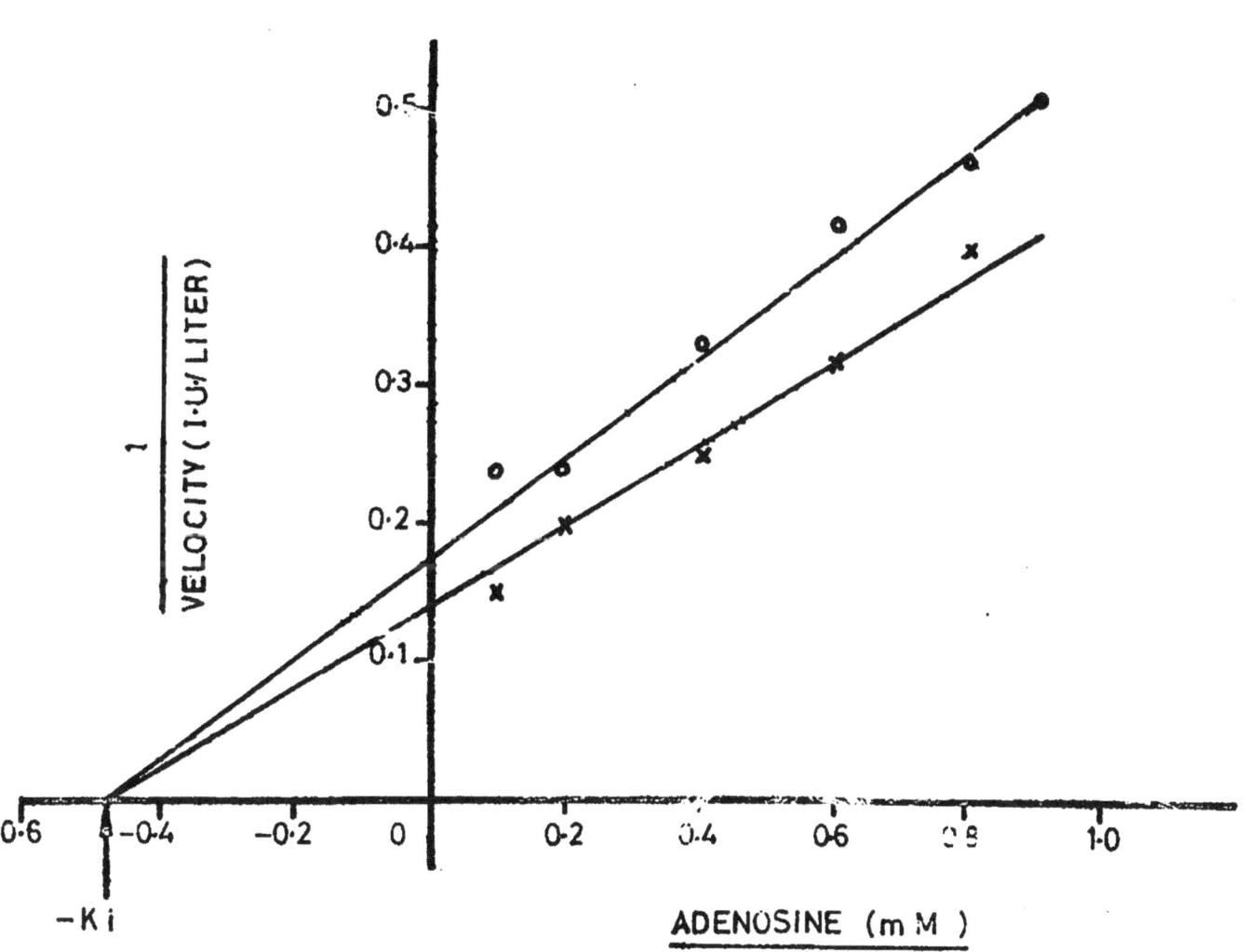

Figure - 18.   The inhibition of 5 - nucleotidase by
adenosine for liver cirrhotic sera.

Details are as in " Figure 17 ".

x ———— x 1.0 mM A - 5 - MP, o ———— o,
0.6 mM   A - 5 - MP.

Fig·18

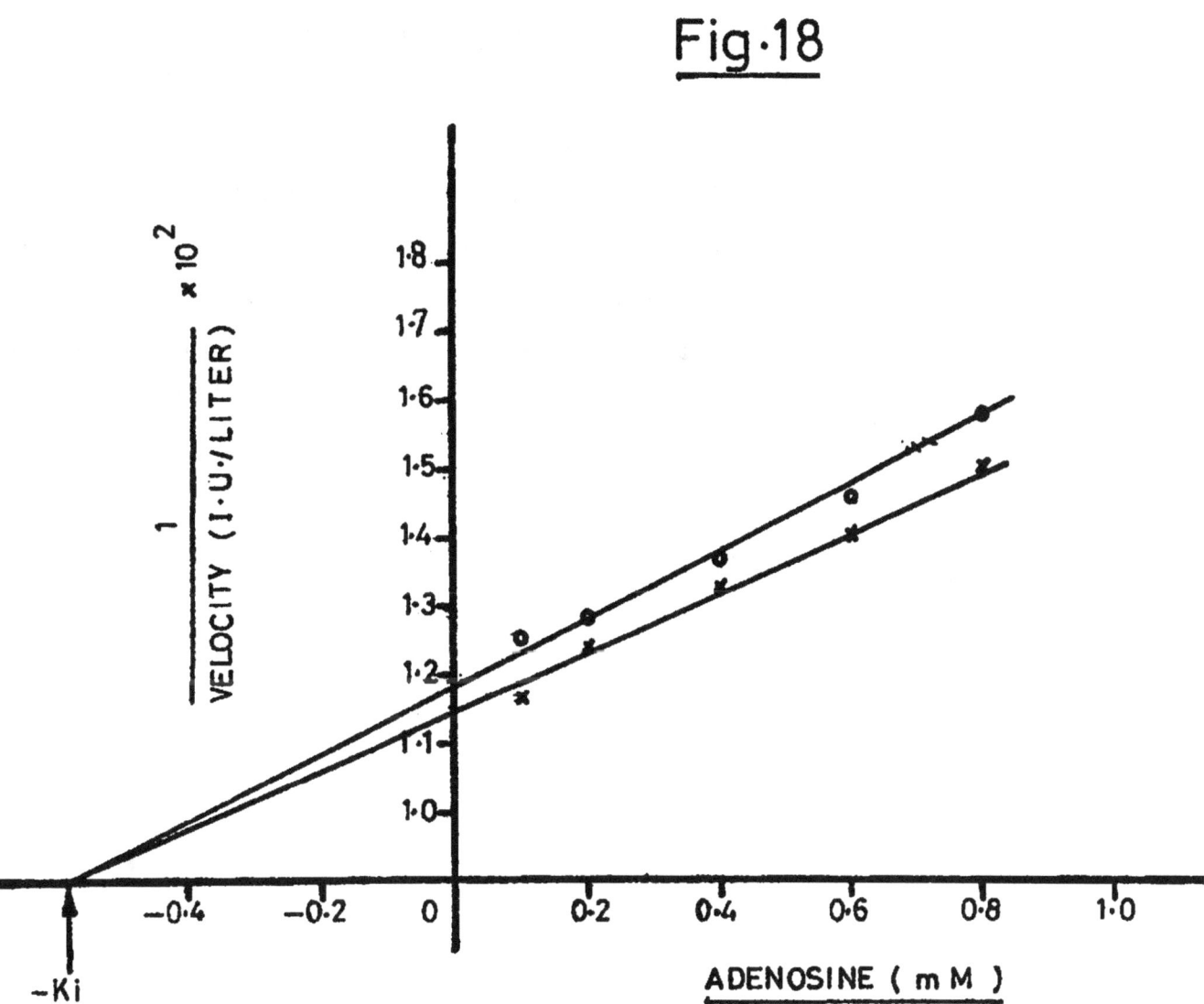

ADENOSINE (m M)

$\dfrac{1}{\text{VELOCITY (I·U·/LITER)}} \times 10^2$

-Ki

Figure - 19. Determination of the degree of inhibition
of 5 - nucleotidase by adenosine for both
normal and liver cirrhotic individuals

The degree of inhibition was determined from
velocity measurements in veronal buffer
pH 7.5 at varying concentrations of adenosine
and at 1.0 mM A - 5 - MP, x ——— x, for
normal serum, o ———- o, for liver cirrhotic
serum.

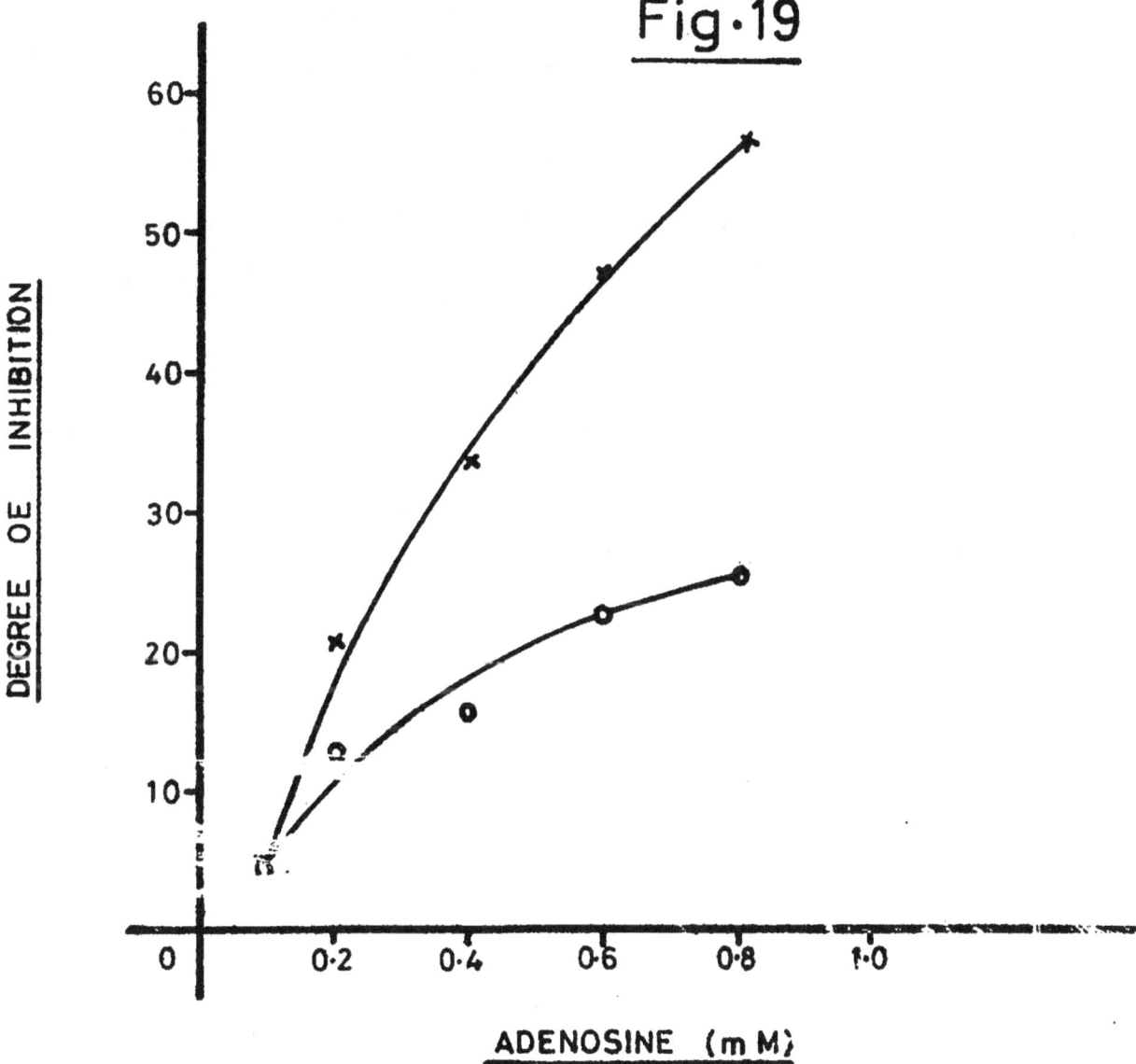

Fig·19

Figure - 20. The plot of velocity for $\bar{5}$ - nucleotidase
against the concentrations of adenosine
for both normal and liver cirrhotic individuals

The rate of the reaction was determined in
the presence of different concentrations of
adenosine, at 1.0 mM A - 5 - MP and pH 7.5

x ———— x (A), for normal serum.
▲ ———— ▲ (B), for liver cirrhotic serum.

Fig·20(A)

Fig·

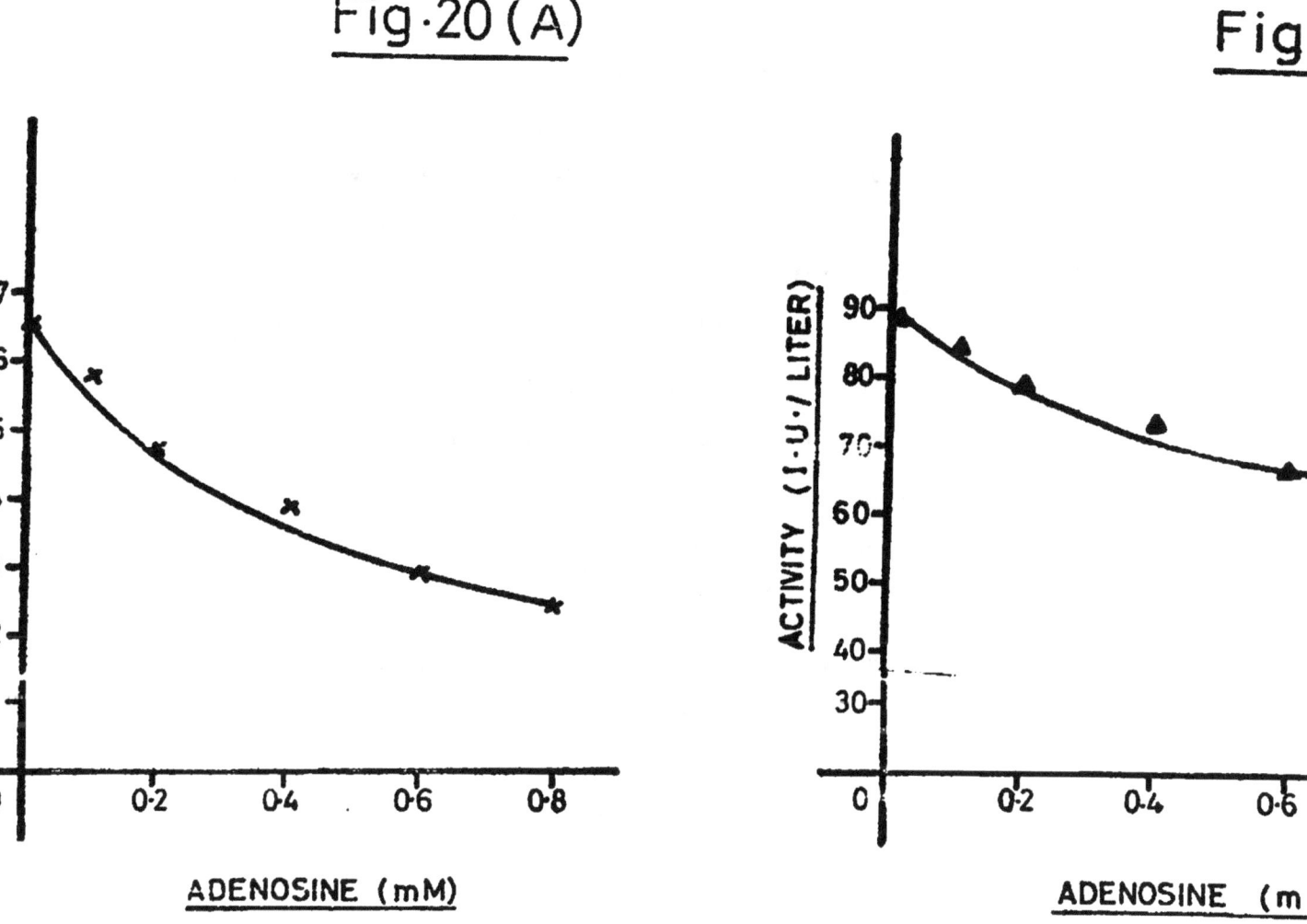

ADENOSINE (mM)

ADENOSINE (m

Figure – 21.   The effect of 20 mM Mg^{+2} on A – 5 – MP – velocity relationship of $\bar{5}$ – nucleotidase in normal individuals

The reaction rate was determined in the presence (o ——— o) and absence (x —·— x) of 20 mM Mg^{+2} using different A – 5 – MP concentrations; 0.1 mM, 0.2 mM, 0.4 mM, 0.6 mM, 0.8 mM, 1.0 mM.   At pH 7.5. The velocity was expressed as I.U./liter.

$\bar{K}m$ = Km (A – 5 – MP) in the presence of Mg^{+2} ions.

Km = Km (A – 5 – MP) in the absence of Mg^{+2} ions.

Fig·21

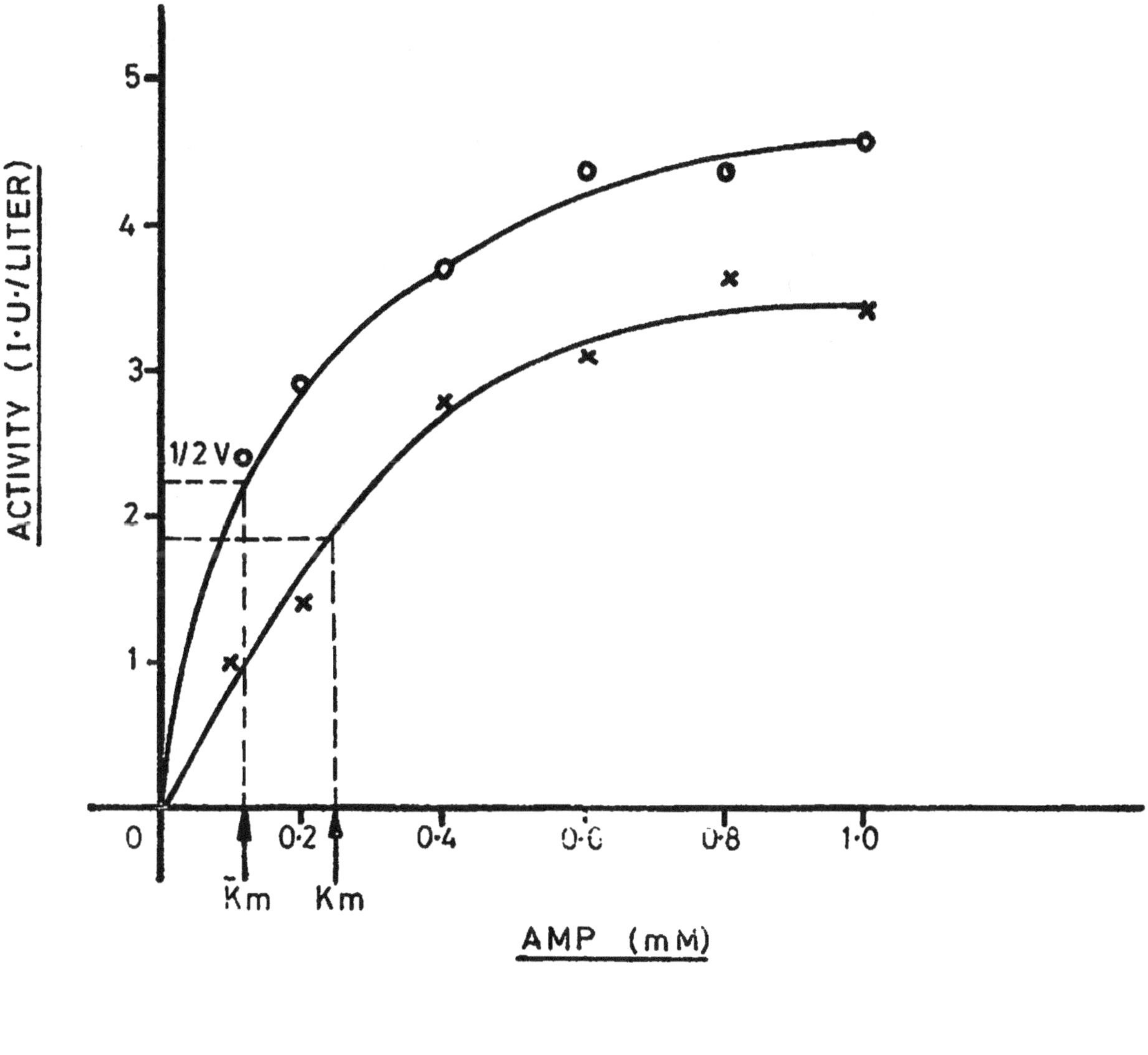

AMP (mM)

Figure - 22.   The effect of 20 mM Mg^{+2} on A - 5 - MP - velocity relationship of 5 - nucleotidase activity in liver cirrhotic sera.

Details are as in " Figure 21 ".

$\bar{K}m = Km$ (A - 5 - MP) in presence of Mg^{+2}

$Km = Km$ (A - 5 - MP) in absence of Mg^{+2}

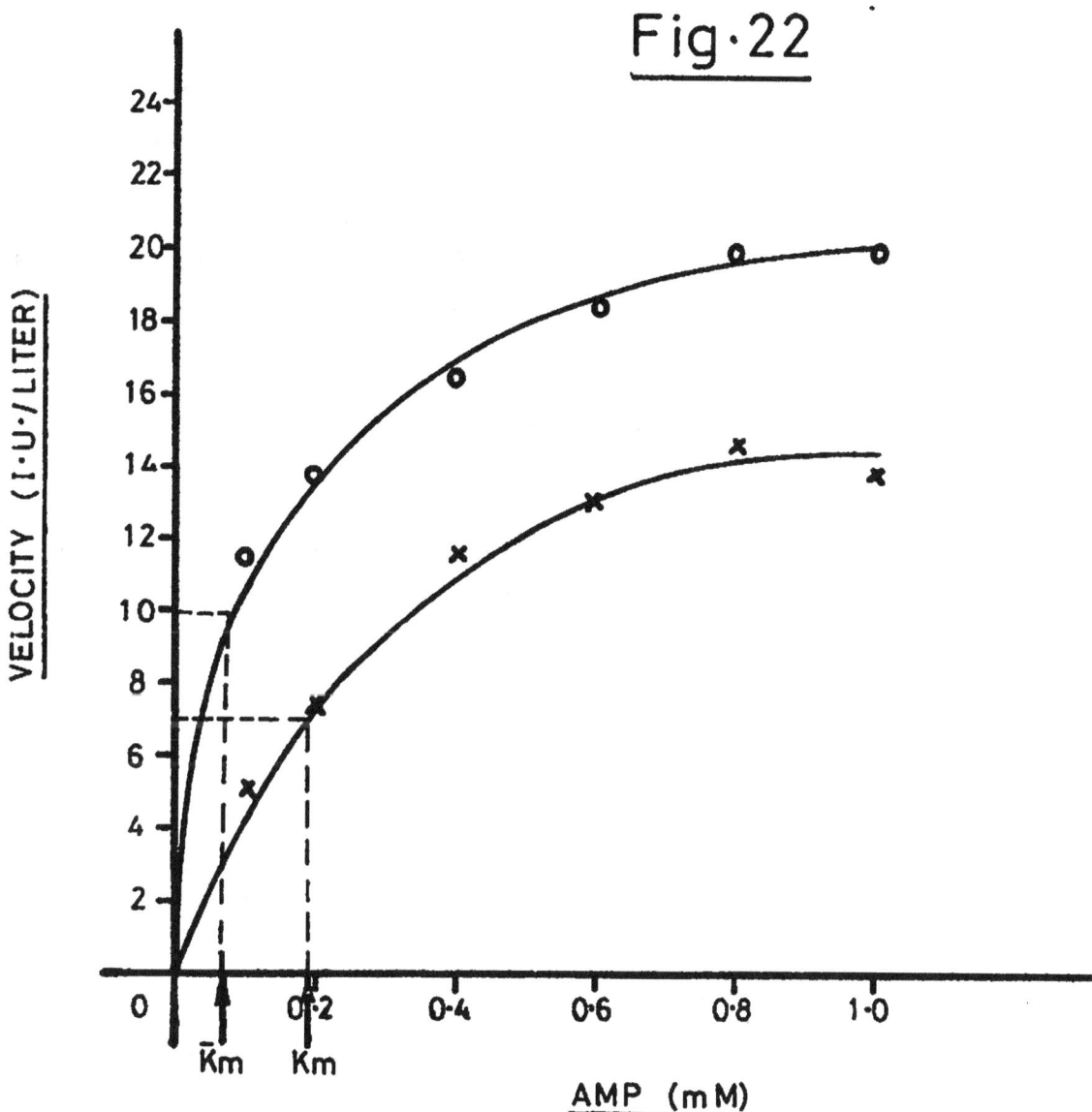

Fig·22

Figure - 23. The effect of 1.0 mM Mn^{+2} on A - 5 - MP - velocity relationship of 5 - nucleotidase activity in normal controls.

The reaction rate was determined in the presence at (x ———— x) and absence (o ———— o), of 1.0 mM Mn^{+2} using different A - 5 - MP concentrations (0.05 mM - 1.0 mM) at pH 7.5.

$\bar{K}m$ = Km (A - 5 - MP) in the presence of Mn^{+2}.

Km = Km (A - 5 - MP) in the absence of Mn^{+2}.

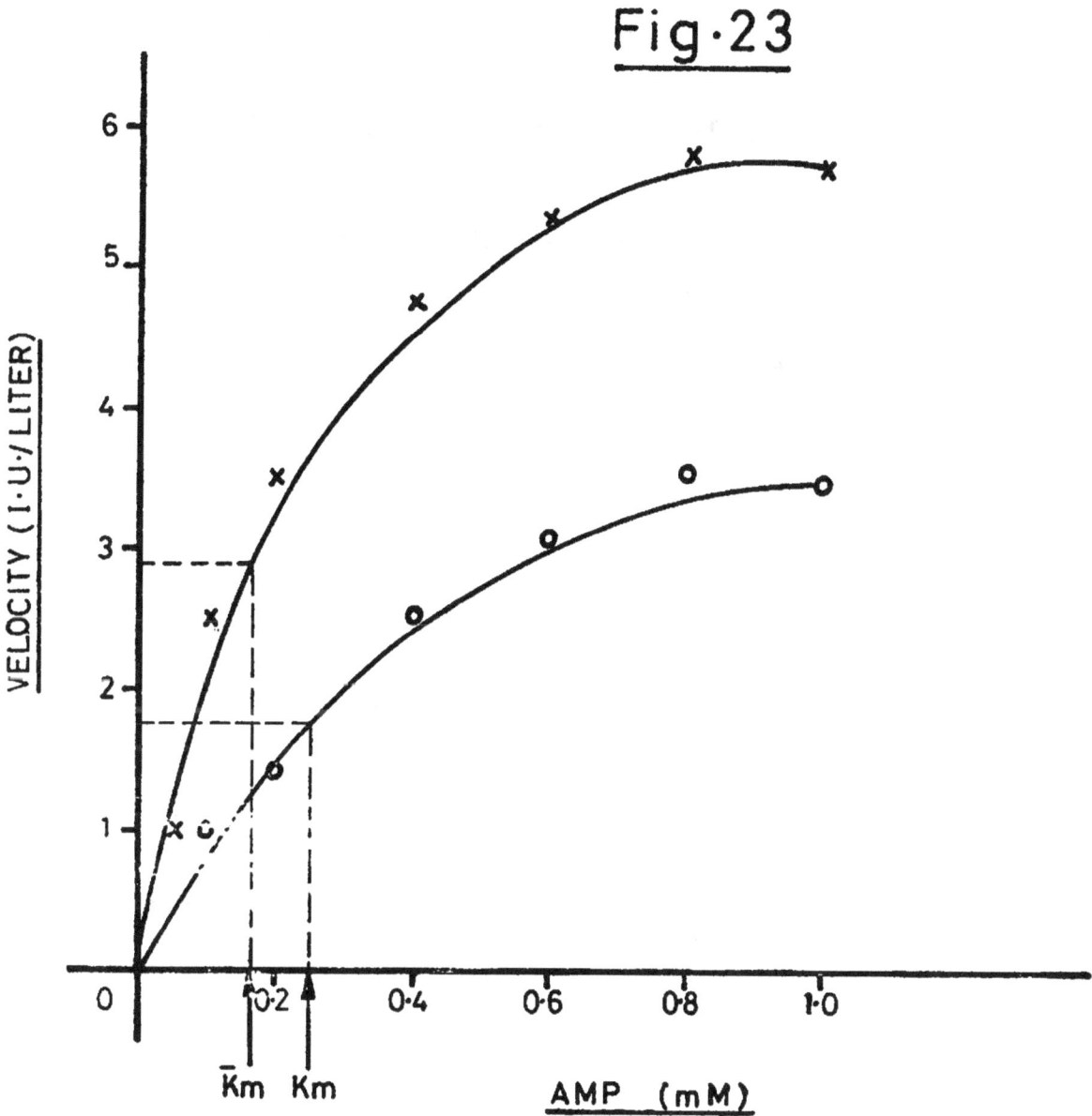

Fig·23

Figure - 24. The effect of 1.0 mM Mn^{+2} on A - 5 - MP - velocity relationship of 5 - nucleotidase activity in liver cirrhotic sera.

Details are as in " Figure 23 ".

$\overline{Km}$ = Km (A - 5 - MP) in the presence of Mn^{+2} (▲ ——————— ▲).

Km = Km (A - 5 - MP) in the absence of Mn^{+2} (x ————— x).

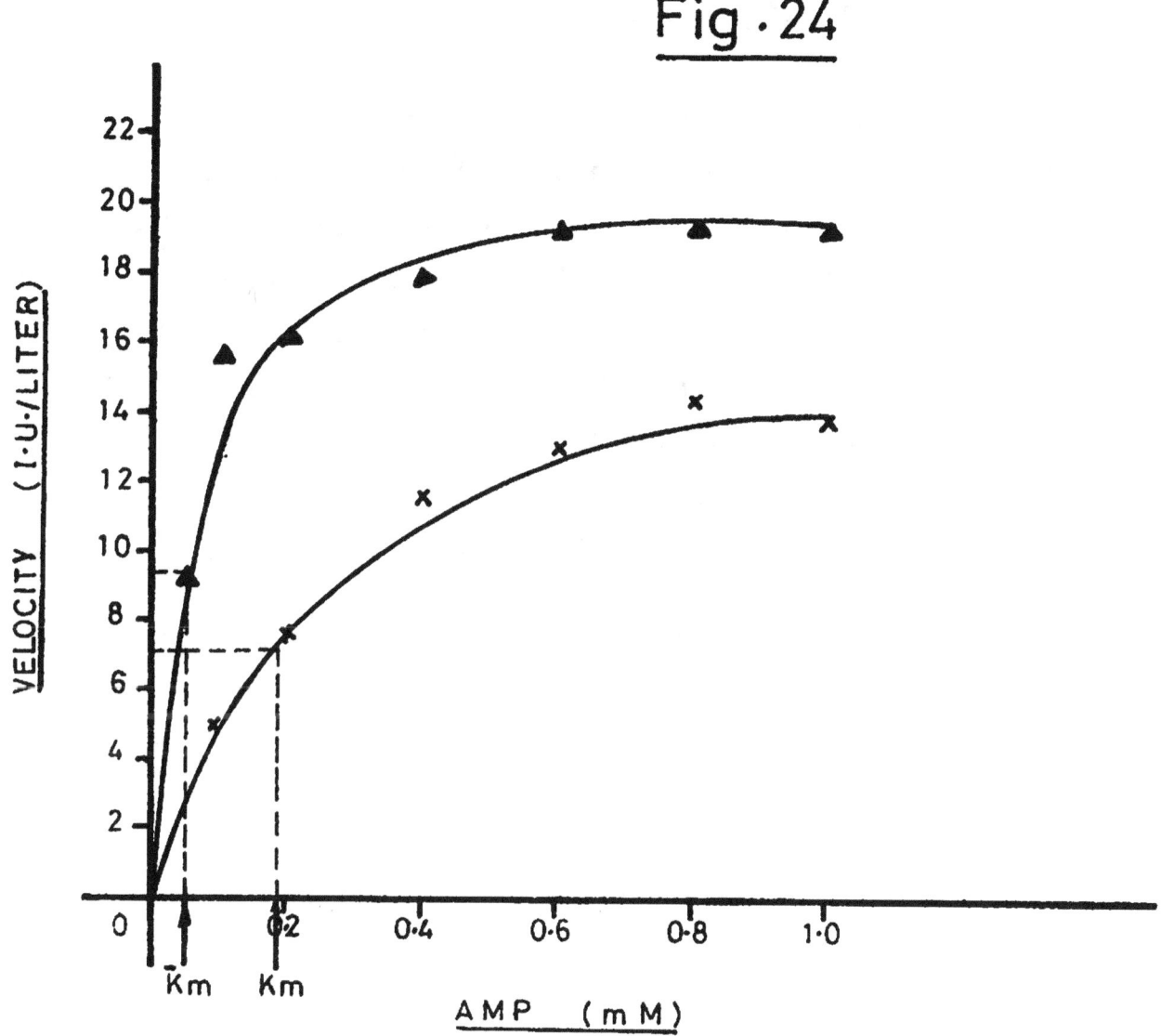

Fig. 24

Figure – 25. <u>The effect of temperature on the reaction rate of $\bar{5}$ – nucleotidase in both normal and liver cirrhotic sera.</u>

The velocity of $\bar{5}$ – nucleotidase was determined at different incubation temperatures for 30 min. at optimum conditions of substrate concentration (1.0 mM   A – 5 – MP) and at pH 7.5.   The temperatures used were; 8°, 25°, 37°, 45°, 60° and 100°. The activity was calculated in I.U./liter.

o ——— o, for liver cirrhotic sera.

x ——— x, for normal sera.

Fig ·25

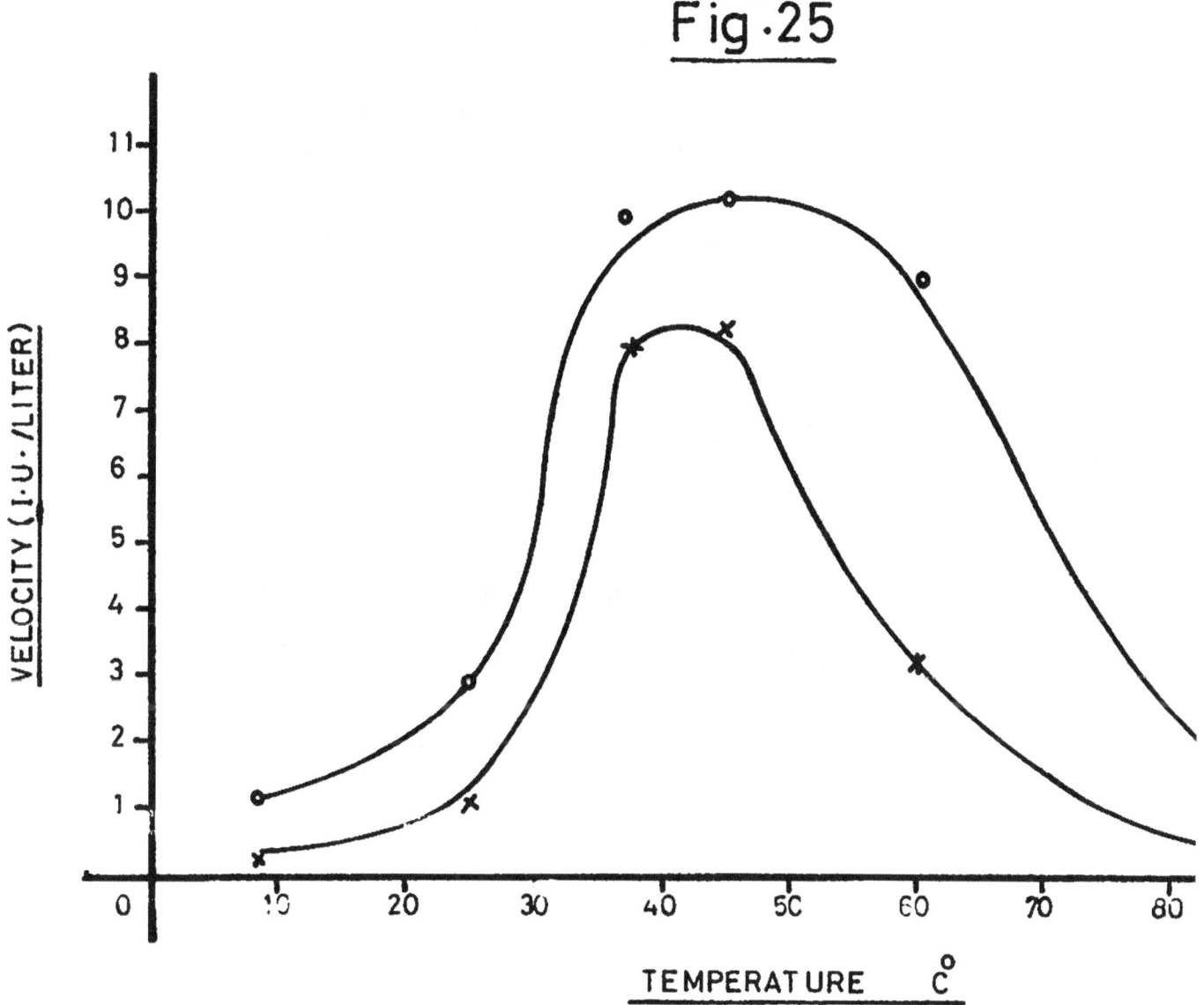

Figure - 26.   The Arrhenius plot, for the effect of
temperature on 5 - nucleotidase activity
in both normal and liver cirrhotic sera.

The integrated Arrhenius[142] equation is

$$\ln k = -\frac{E}{R} \left(-\frac{1}{T} - \right) + \text{constant}$$

E = activation energy

k = rate constant

T = absolute temperature

R = gas constant

A plot of log V versus 1/T give a straight
line corresponding to this figure.

o ——— o, for pathological sera

x ——— x, for normal sera.

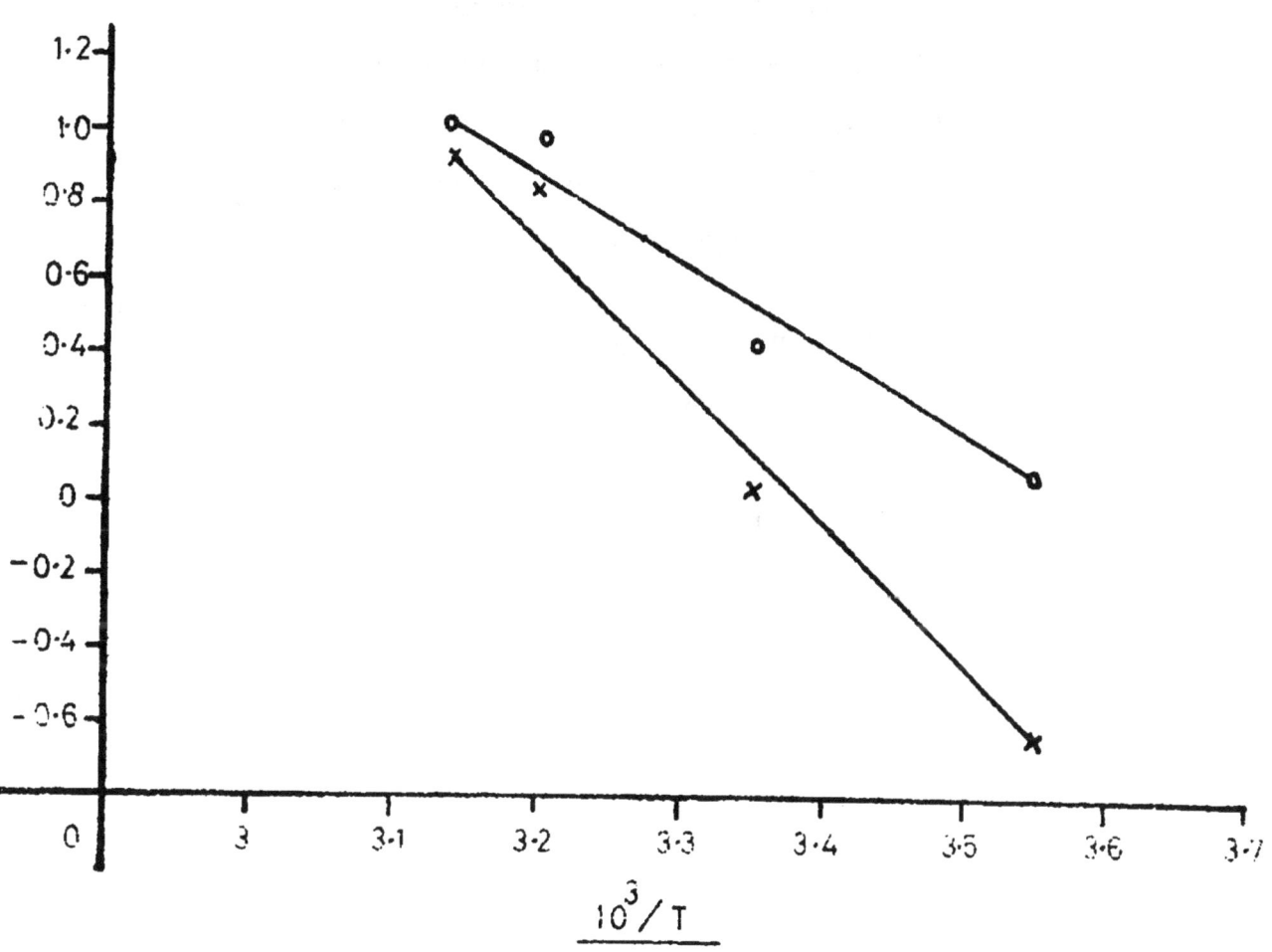

Fig ·26

$10^3/T$

Figure – 27.   pH – velocity relationship of 5 – nucleotidase
at different A – 5 – MP concentrations for
normal and cirrhotic sera.

The velocity of the reaction was measured
at different A – 5 – MP concentrations in
the pH range (7 – 8.5), using veronal
buffer at 37°. Velocity measurements were
expressed in I.U./liter. The A – 5 – MP
velocity relationship are given at these
A – 5 – MP concentrations,
0.2 mM x ——— x, 0.4 mM, o ——— o,
0.8 mM . ——— ., for normal (A) and
liver cirrhotic (B) individuals.

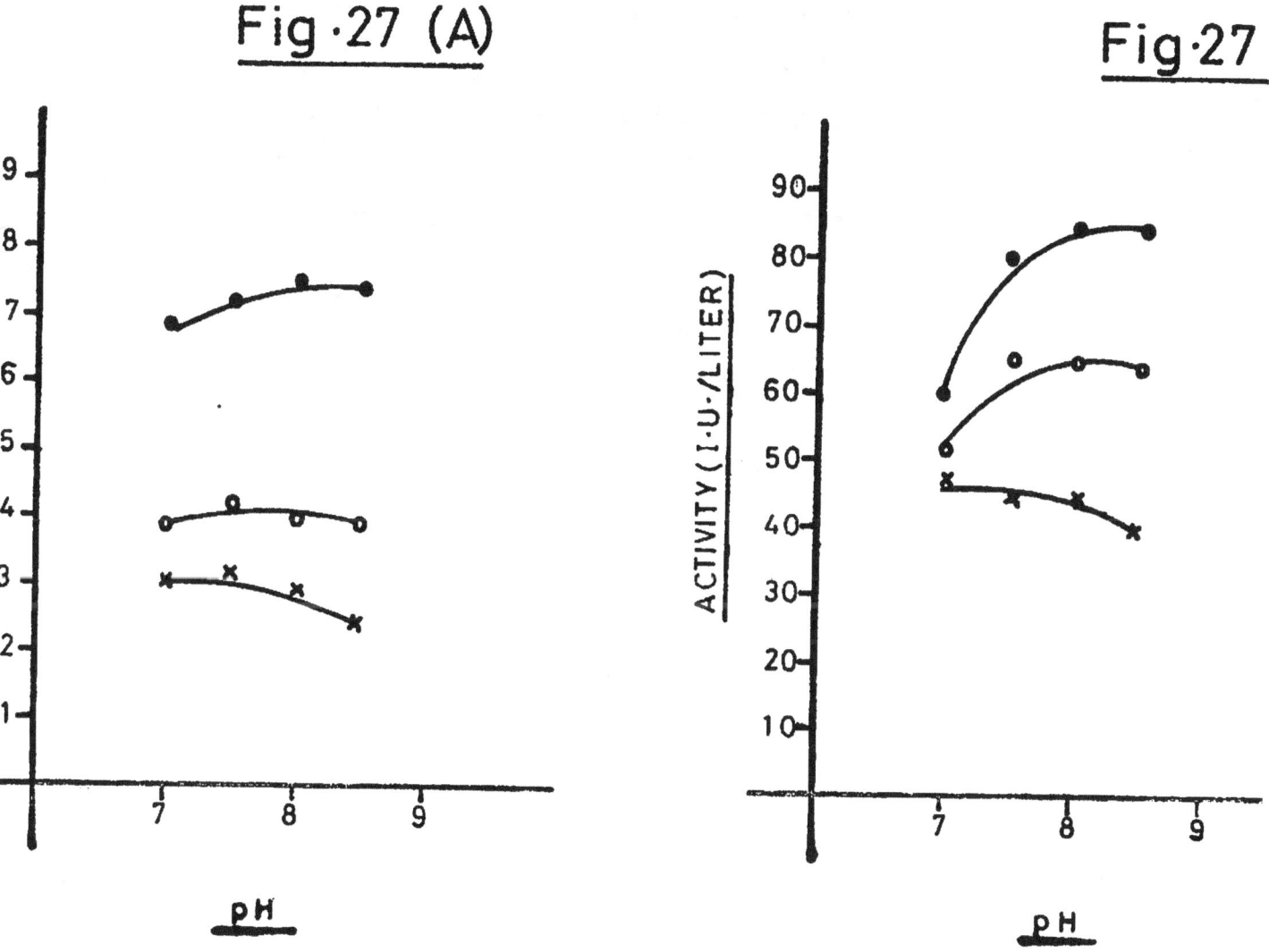

Fig·27 (A)

Fig·27

150

Figure - 28. Variation of optimum substrate concentrations
(A - 5 - MP) of 5 - nucleotidase with pH
in normal and liver cirrhotic individuals.

Velocity of the reaction was measured at
different pH values (7, 7.5, 8, 8.5),
using veronal buffer and different A - 5 - MP
concentration (0.1 - 1.2 mM) at 37°.
Details are as in " Experimental ".

x ———— x, for normal, o ———— o, for
liver cirrhotic sera.

Fig · 28

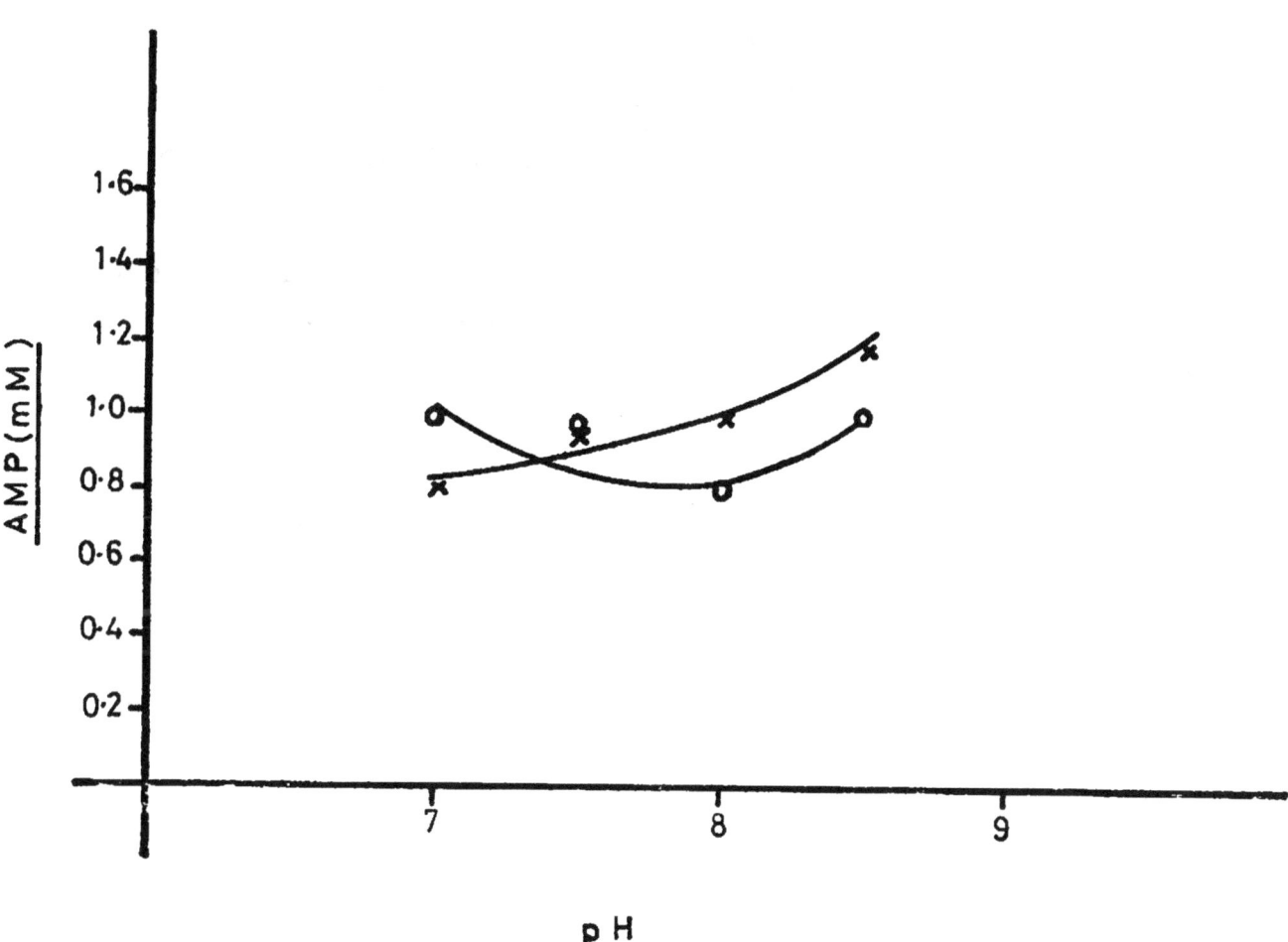

Figure – 29. The effect of pH on the initial velocity of
the reaction for $\bar{5}$ – nucleotidase in normal
and liver cirrhotic sera.

The reaction was carried out at different
pH values (7 – 8.5) using veronal buffer.
Initial velocities were determined at 37°
in the presence of 0.4 mM A – 5 – MP.

x —— x, for liver cirrhotic patient
o —— o, for a normal individual

Fig · 29

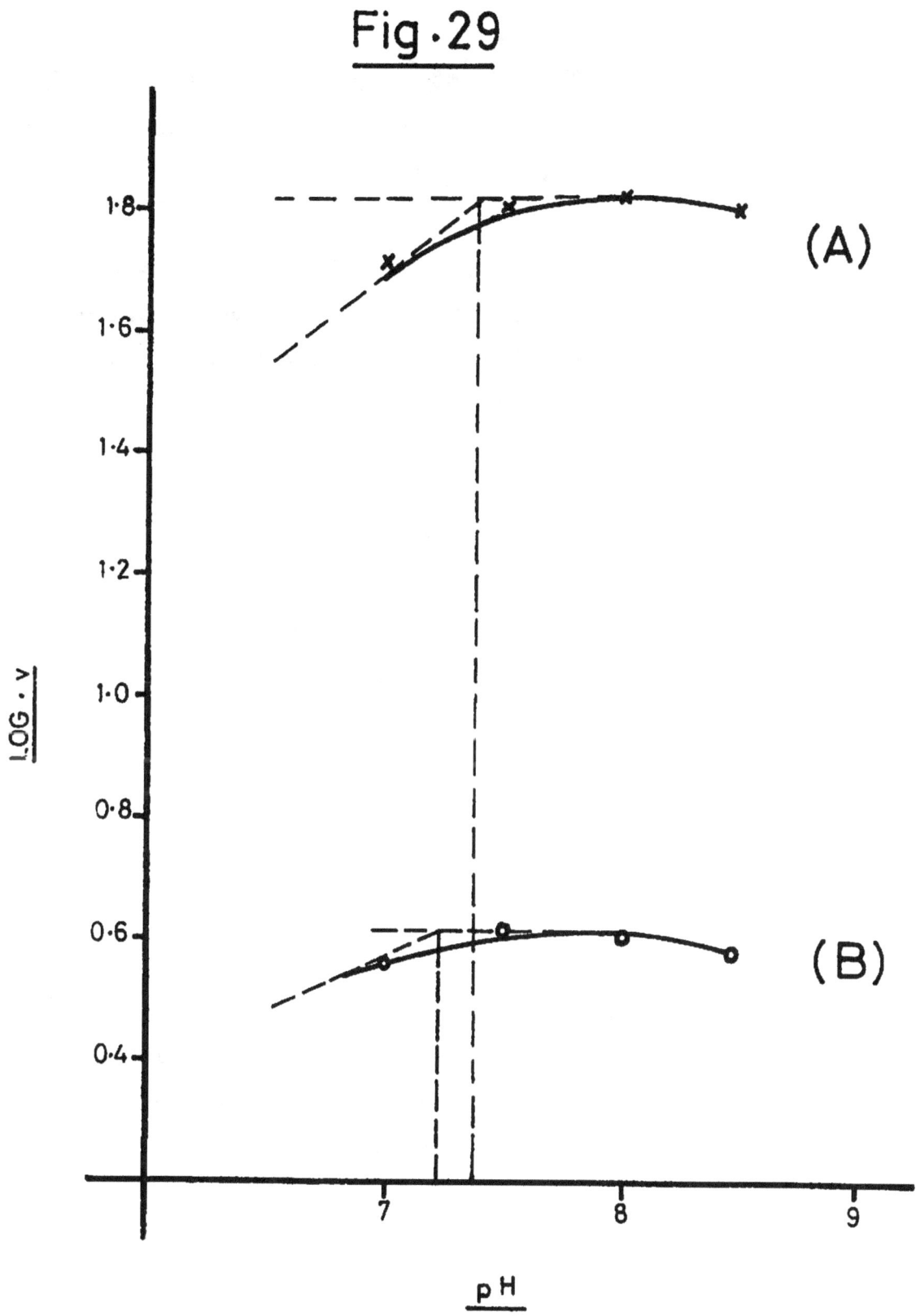

Figure - 30.   The effect of pH on Km (A - 5 - MP)[*] for
5 - nucleotidase in a normal and liver
cirrhotic patient.

Velocity measurements were taken at different
A - 5 - MP concentrations (0.1 - 1.2 mM),
at pH values ranging from (7 - 8.5), using
veronal buffer, at 37°.

o ———— o for normal,   x ———— x   for
liver cirrhotic.

---

[*] Km (A - 5 - MP) was determined from a
plot of v versus (S) according to the
original Michaelis - Menton equation.

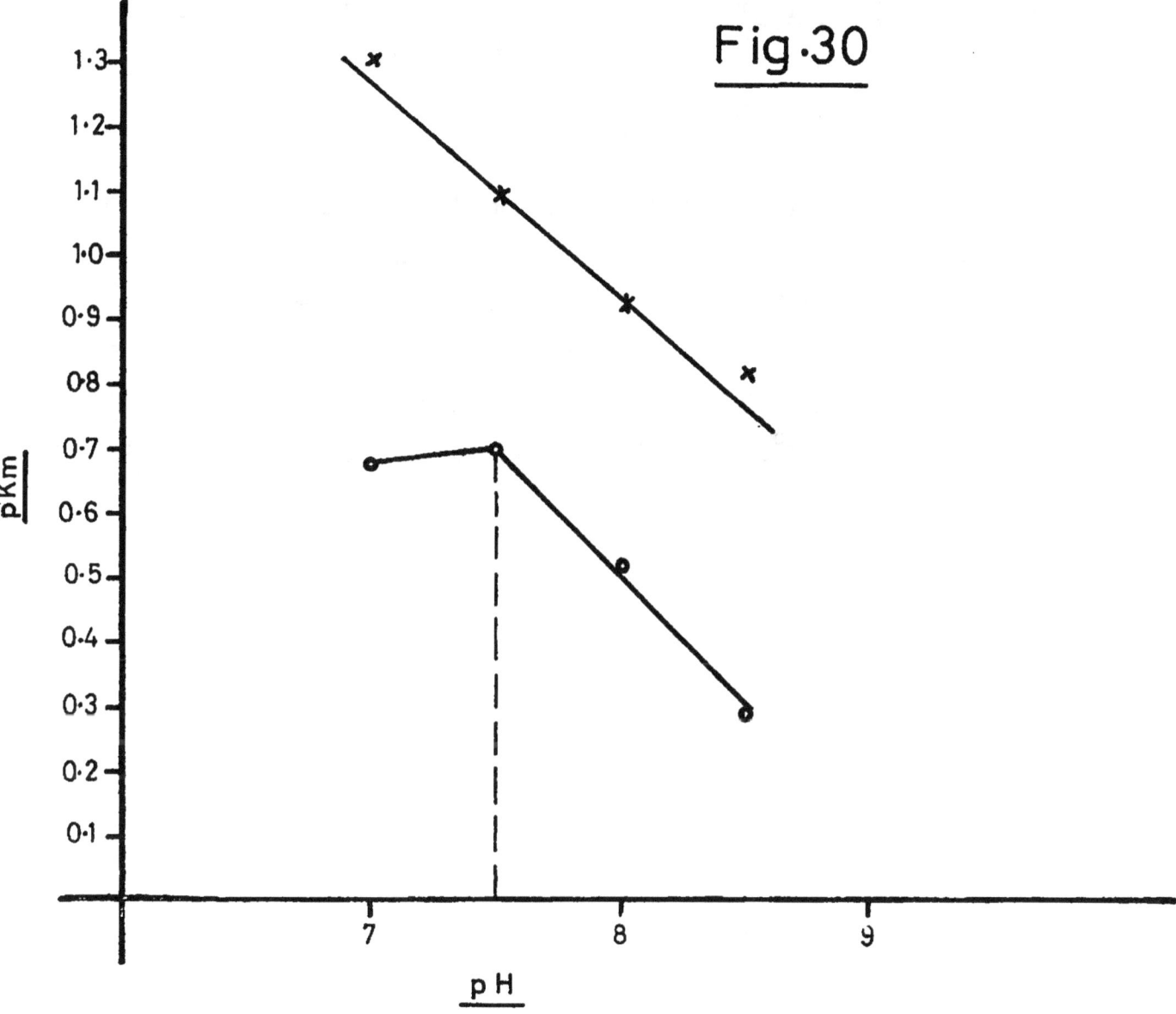

Fig.30

Figure – 31. pH optimum determination, in the presence
and absence of 20 mM Mg^{+2} ions, at 37°, for
normal persons.

The reaction rate was determined at 1.0 mM
A – 5 – MP , and varying pH (7 – 8.5)
using veronal buffer, in the presence
(o ———— o) and absence (x ———— x) of
20 mM Mg^{+2}.   Measurements were taken at
37°. Details are as under " Experimental ".

Fig·31

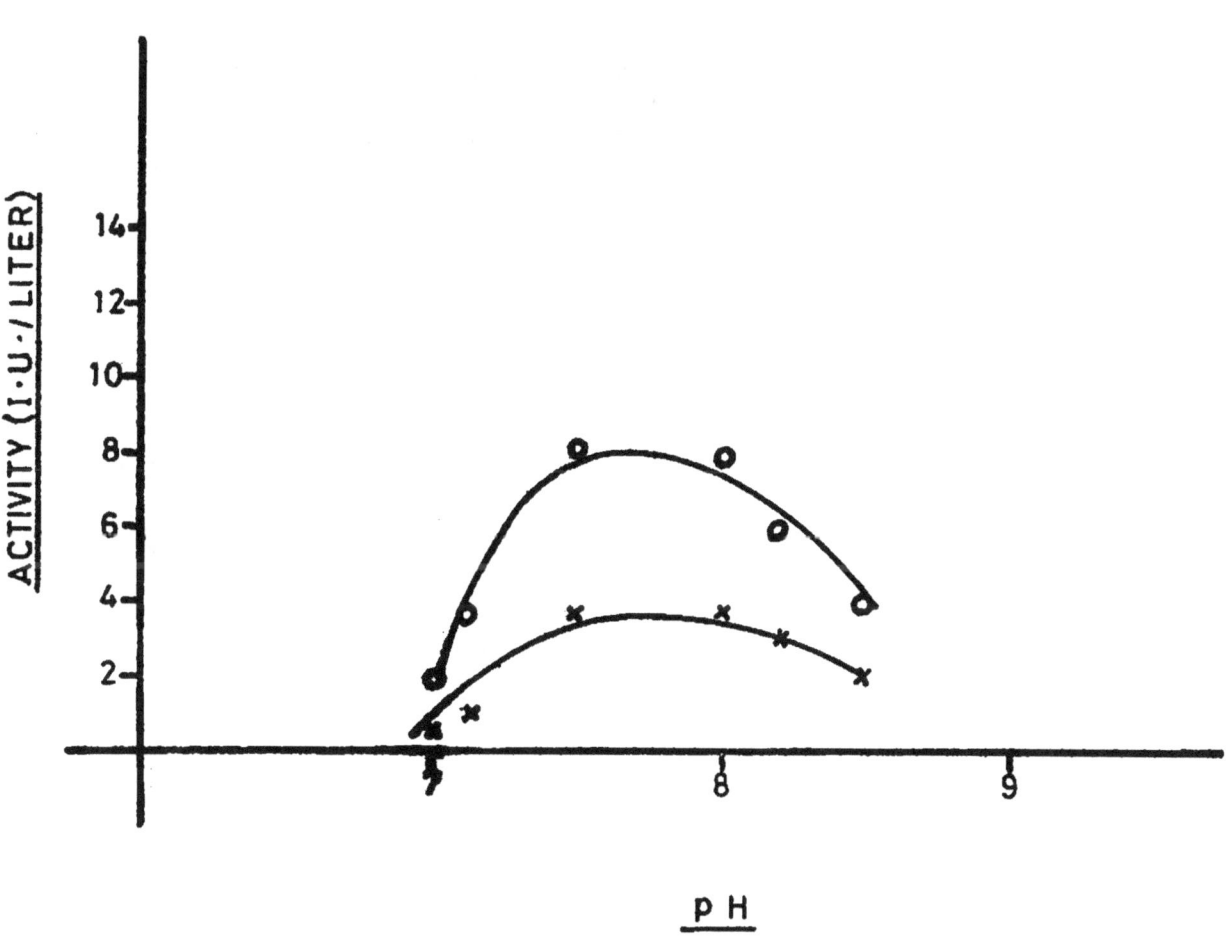

Figure - 32. pH optimum determination in the presence and absence of 20 mM Mg^{+2} ions, at 37°, for patients with liver cirrhosis .

Details are as in " Figure 31 ".

x ———— x, in the absence of Mg^{+2}
o ———— o, in the presence of Mg^{+2}.

Fig·32

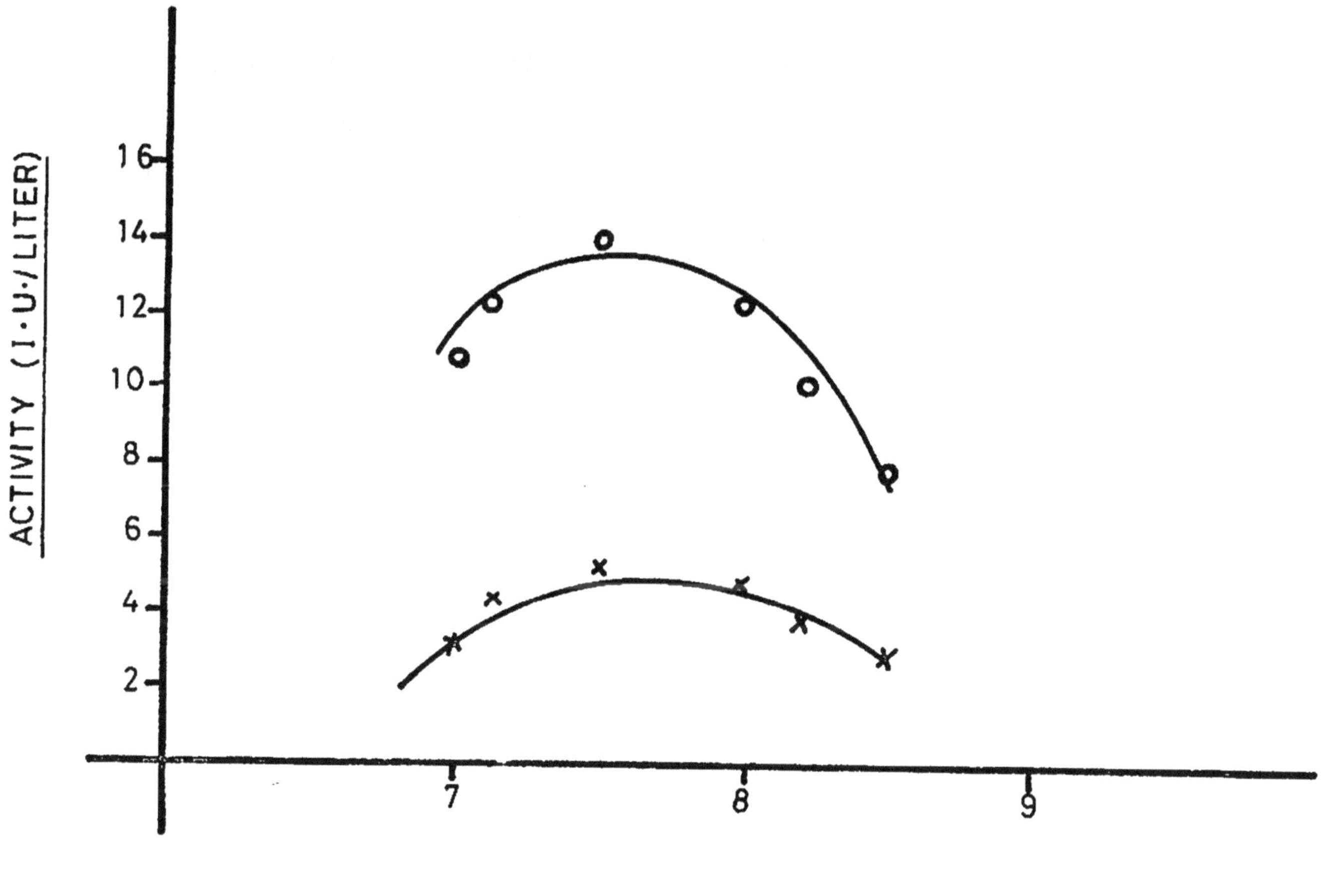

ACTIVITY (I·U·/LITER)

16 –
14 –
12 –
10 –
8 –
6 –
4 –
2 –

7   8   9

pH

Figure – 33. The plot of $\log \dfrac{V}{V - v}$ against log (A–5–MP), for normal persons.

The interaction coefficient n, was determined for normal sera in the absence of inhibitor. The velocity was measured at 37°, and pH 7.5 using veronal buffer. The A – 5 – MP concentrations used ranged from 0.05 to 1.0mM.

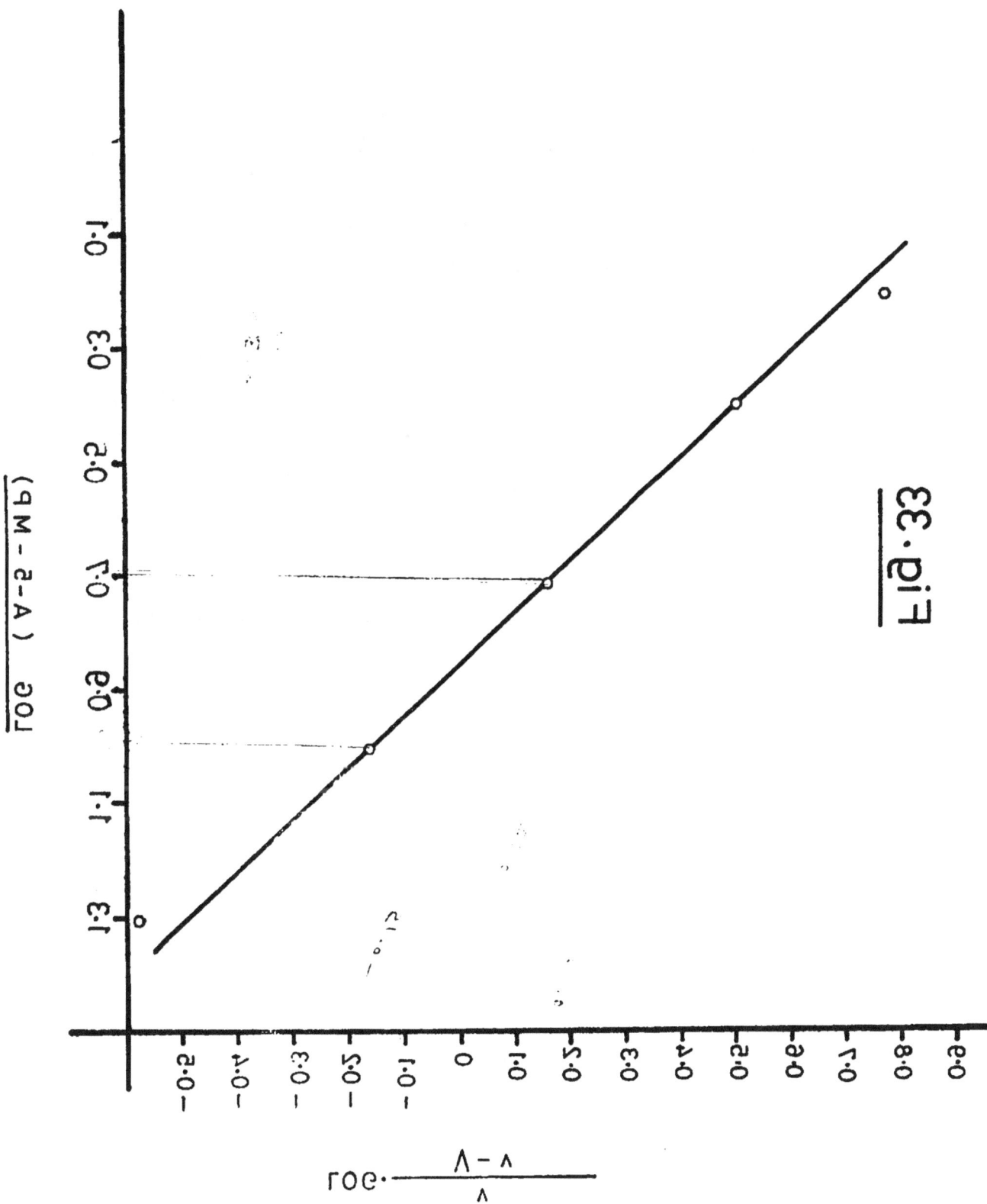

Fig·33

Figure - 34. The plot of log $\dfrac{v_i - v_i \ (\text{Sat})}{v_o - v_i}$ against log (ATP), for both normal and liver cirrhotic sera.

The reaction rate was determined in the presence of different concentrations of ATP, as described in point (11) of the " Experimental ", at 37°.

▲ ———— ▲ , for normal, o —————— o, for liver cirrhotic sera, at 0.6 mM, A - 5 - MP.

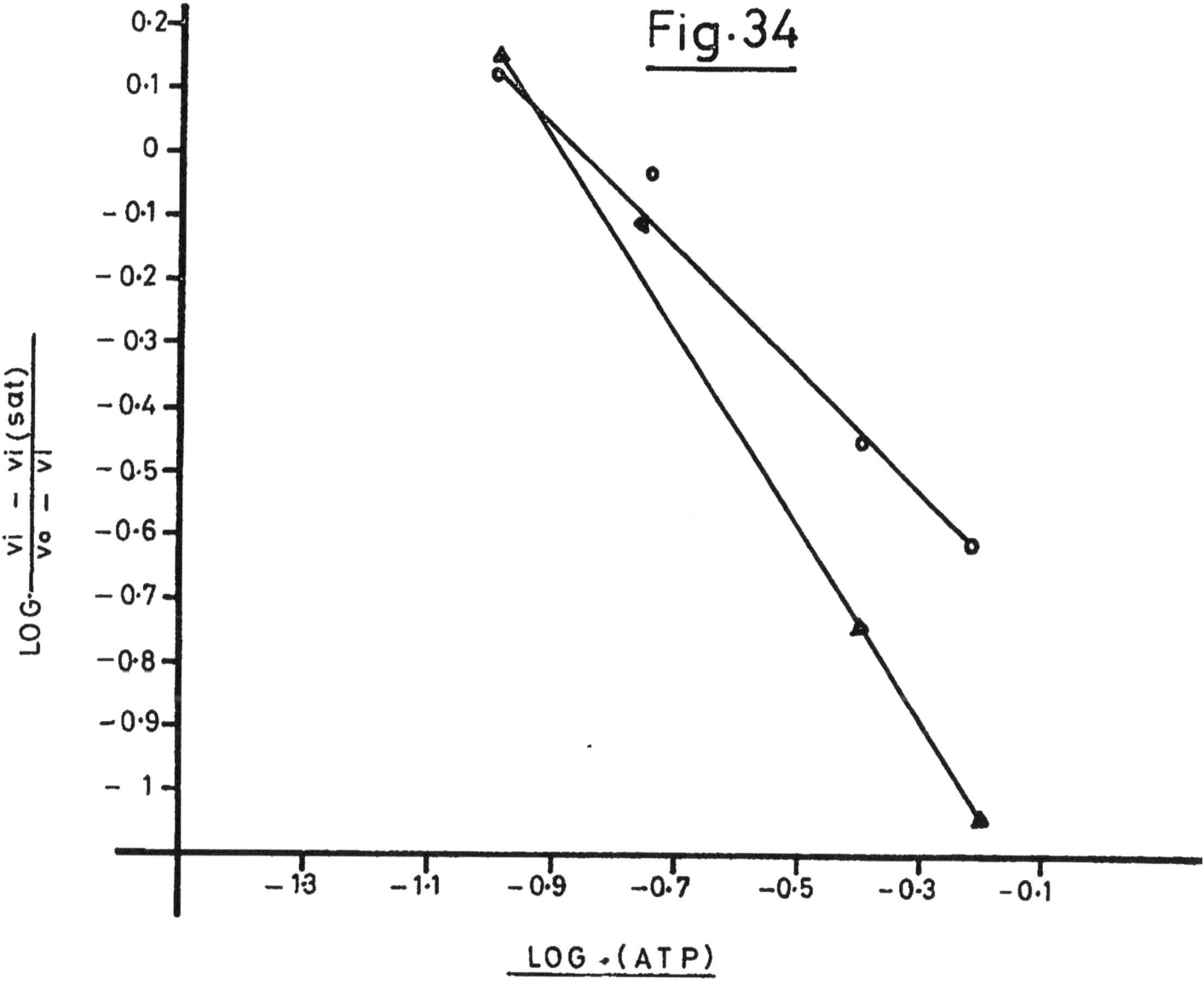

Fig. 34

CHAPTER FOUR
DISCUSSION

D I S C U S S I O N

Blood normally contains those enzymes which have their origin in the cells of various tissues of the body, if the cells remained intact indefinitely, the cellular enzymes could probably never escape into the blood stream, however, under some clinical conditions, in the case of acute considerable damage to enzyme - rich organs, cellular enzymes leak into the interstitial fluid and find their way into the blood, therefore the determination of enzymes levels in blood, allows the easy detection of the origin of that pathological condition, and aids in the diagnosis and prognosis.

Determination of various enzymes, and measurement of different parameters of enzymatic reactions, are considered a major diagnostic tool[20] for many diseases. Various workers have determined these parameters under different conditions the significance of these determinations will probably clarify a great deal of enzymatic behavior in these diseases [143 - 146].

Enzymes are important and essential components of biological systems, their function being to catalyze the

Biochemical reactions, without the efficient aid of the
enzymes, these processes would occur at greatly diminished
rates, or not at all.

This work is mainly concerned with the kinetics of
the reactions catalyzed by $\bar{5}$ - nucleotidase in both normal
Iraqi and liver cirrhotic individuals, the results of the
investigation on $\bar{5}$ - nucleotidase in liver cirrhotic and
normal sera provide better picture for a mammalian
$\ddot{5}$ - nucleotidase than has been thus far available. The
catabolism of purine nucleotides and pyrimidine nucleotides
starts with the action of $\bar{5}$ - nucleotidase. All $\ddot{5}$ - nucleotides are hydrolyzed by the same enzyme or by different
enzymes localized in the same or different cells. Since
$\bar{5}$ - nucleotidase catalyze, the 1st step in the catabolism
it is hardly conceivable that in cells showing a high
$\bar{5}$ - nucleotidase activity, synthesis of nucleic acid can
take place, so the $\bar{5}$ - nucleotidase could be concerned
with the control of nucleic acid synthesis. The metabolism of the purine nucleotides[147] has been summarized
in Scheme I and II. The anabolic reactions are indicated
by straight lines and the catabolic ones by broken lines.

Scheme I. Metabolism of purine nucleotides

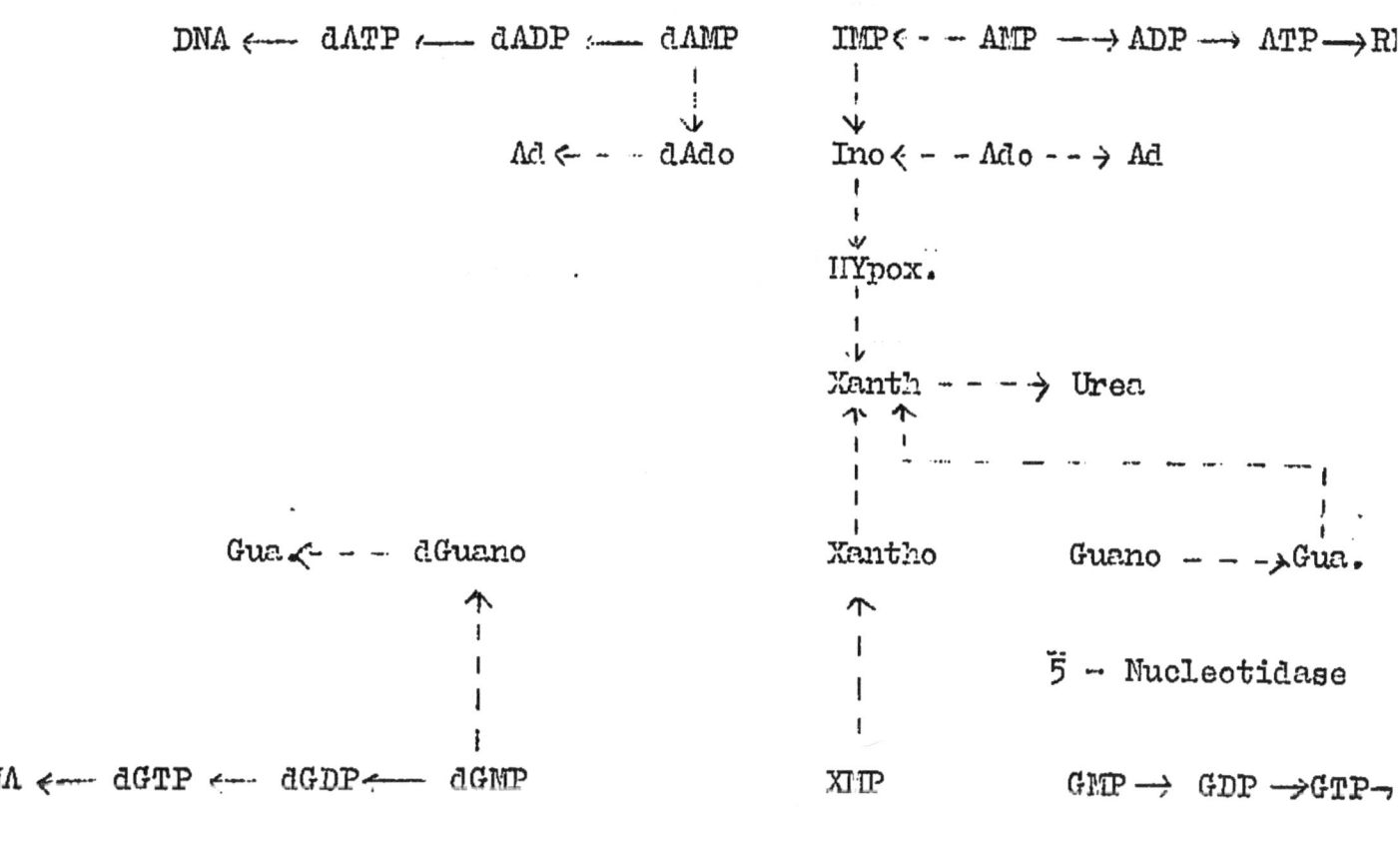

DNA ⟵ dATP ⟵ dADP ⟵ dAMP   IMP ⟵ - - AMP ⟶ ADP ⟶ ATP ⟶ RI

                   Ad ⟵ - - dAdo   Ino ⟵ - - Ado - - ⟶ Ad

                                 HYpox.

                                 Xanth - - - ⟶ Urea

Gua ⟵ - - dGuano   Xantho   Guano - - - ⟶ Gua.

                                           5 - Nucleotidase

DNA ⟵ dGTP ⟵ dGDP ⟵ dGMP   XIP   GMP ⟶ GDP ⟶ GTP ⟶

                    —————— = Anabolic processes.

                    —·—·—·— = Catabolic processes.

Scheme II. Catabolism of Nucleic Acids

DNA (RNA)

| DN-ase (RN - ase)
↓

Polynucleotides

Phospho diesterase
↓

5 - Nucleotides (3 - Nucleotides)

5 - Nucleotidase
↓

Nucleosides

↓

Urea

5 - nucleotidase as a catalyst, it influences the rate of the following chemical reaction, but it is not itself used up during the process and can in ideal cases be recovered at the end of the reaction.

$$A - 5 - MP \xrightarrow[\text{5 - nucleotidase}]{K_1} \text{Adenosine} + \text{Pi} \qquad (1)$$

the rate of this reaction may be given by

$$v = K_1 \ (\ E\ ) \ (\ A - 5 - MP\ ) \qquad (2)$$

Besides all of the above respects the enzyme 5 - nucleotidase exhibit the typical properties of catalysts, it shows unusual characteristics which are not found with other types of catalysts, among these are those properties, which concern sensitivity to heat and extremes of pH, in which it shows the typical behavior of proteins.

5 - Nucleotidase function through the formation of an 5 - nucleotidase - A - 5 - MP intermediate complex, and initial velocity obtained is often refered to as the Michaelis - Menten equation (Figure 3)

$$v_o = \frac{K_{So}}{1 + S_o / K_s} \qquad (3)$$

Where v_o and S_o are the velocity of the enzyme
- catalyzed reaction and substrate concentration at zero
time.

Graphical methods for evaluating the constants of
the rate equation (V and Kn) of Michaelis - Menten
and extended work, could be applied for 5 - nucleotidase,
the concept of an enzyme - substrate intermediates complex
often referred to as a " Michaelis Complex " as the
theoretical basis of enzyme kinetics has proved extremly
fruitfull (Figure 4), extentions of the theory have
dealt with the effects of competitive and non - competitive
inhibitors, activators, have been applied to 5- nucleotidase
(from Figure 10 to 24).

Activity measurements of 5 - nucleotidase

Although 5 - nucleotidase, is widely distributed
in body tissues, an increase in its serum level appears
to occur only in liver diseases[96], due to the abnormali-
ties in the liver, caused by the increased activity in
enzyme synthesis of the cells which produce this enzyme,
or of proliferation of these cells, or both. If the
cell membrane becomes permeable, active retention of the
enzyme 5 - nucleotidase gradually ceases, and it appears
in the extracellular space.

169

The main value of $\overline{5}$ - nucleotidase over the non-specific AP measurements in serum, is in deciding if a raised serum AP level is due to bone or hepato - biliary diseases, as most studies show that serum $\overline{5}$ - nucleotidase does'nt rise in bone disease[33, 97], but AP rises in this disease.

In the present studies of 40 patients with liver cirrhosis, the activity of $\overline{5}$ - nucleotidase ranged between 5 and 50 I.U./liter which was higher than that of normal Iraqi individuals (1.45 - 12.09 I.U./liter), and the percent increase in activity over normal was found to be 57.37%. In 8 cases out of 40 patients with liver cirrhosis showed a great increase in activity (20 - 50 I.U./liter), and the percent increase in activity over normal was found to be 76.19% for these 8 cases, which were suggested to be of the biliary cirrhosis (Table - 1).

Soochi et al.[123] reported that raised levels of $\overline{5}$ - nucleotidase occur in serum of patients with liver cirrhosis, which supports our results. The clinical observations by Young[97] suggested that the liver is the most likely source of serum $\overline{5}$ - nucleotidase elevation in hepatobiliary disease. The change in level of $\overline{5}$ - nucleotidase may be due to alterations in the extent to which a type of cell is capable of producing the enzyme or, increase in the number of cells. Greater elevations of $\bar{5}$ - nucleotidase activity were seen more often in patients with biliary cirrhosis, many reports[148] suggest that $\bar{5}$ - nucleotidase originates also from the cells of the bile duct or ductules. This argument hold for our results obtained for the 8 cases.

Table - 2., shows that the activity range do not appears to be influenced by age, or sex in normal Iraqi individuals, in the age groups 20 - 30 and 30 - 40 years, this is supported by the study of Young[97] on 30 normal adults. In the case of liver cirrhosis, patients in the age group from 10 - 20 year, the activity for male was higher than female, but for age group 20 - 30 and 30 - 60 year, there was no marked difference in $\bar{5}$ - nucleotidase activity

Kinetics of $\bar{5}$ - nucleotidase in normal and liver cirrhotic sera

Most of the studies on serum $\bar{5}$ - nucleotidase were previously directed towards activity determination in liver diseases[96, 97, 123].

Due to the specificty of $\bar{5}$ - nucleotidase for liver
diseases, and its important role in the catabolism of purine
nucleotides and pyrimidine nucleotides, it was intended
to study the characteristics of $\bar{5}$ - nucleotidase in liver
cirrhotic and normal Iraqi individuals, to find whether
the changes brought a bout by this disease are of quanti-.
tative or qualitative nature, through the kinetic studies.
The present research gives an account of a few aspects
of general chemical kinetics that are particularly relevant
to $\bar{5}$ - nucleotidase kinetics in both systems.

Stability of $\bar{5}$ - nucleotidase in both normal and liver
cirrhotic sera.

Through the present investigation, the stability
of $\ddot{5}$ - nucleotidase, at different temperatures ranging
from 27° to 60°, for 30 minutes, was studied. Table - 3
reveals that $\bar{5}$ - nucolotidase from liver cirrhotic sera
was less stable than that of normal sera, at the temperatures indicated above. Many reports have attributed
this changes to the distribution of tertiary structure
of protein (denaturation), since the biological activity is very clossely dependent on the correct tertiary
structure of proteins.

Time studies

A conventional kinetic study is applied to $\bar{5}$ - nucleo-tidase, which consist of mixing the reactants (A -- 5 - MP and $\bar{5}$ - nucleotidase) and following the concentration of inorganic phosphate (product) as a function of time; this is the static method.

As a result of complexities of this kind, investi-gations of the time course of enzyme reactions frequently can't be interpreted unambiquously. For example, simple first - order behavior in the course of a one - substrate reaction might be due to either of the following two cases;

1- The substrate concentration might be low, so that the true order is first, or

2- The true order might be zero, but the reaction might be slowed down during a run owing to the inhibition by product.

The reaction which is catalyzed by $\bar{5}$ - nucleotidase involve only one substrate (A - 5 - MP), the rate varies with the substrate concentration according to the Michaelis - Menten Law, but with some differences in feature of the curves in both sources (normal and liver cirrhotic)

$$v = \frac{V\,(S)}{Kn + (S)} \tag{4}$$

See Figure - 3, for both normal and liver cirrhotic individuals.

If so is the initial amount of A - 5 - MP, and X the amount of product (inorganic phosphate) after time t, the law may be written as

$$\frac{dX}{dt} = \frac{V(S_o - x)}{Kn + S_o - x} \tag{5}$$

At high substrate concentration the enzyme becomes saturated with the substrate, and the rate equation reduced to

$$\frac{dX}{dt} = V \tag{6}$$

$$V = \frac{x}{t} \tag{7}$$

A plot of x against t will give a straight line of slope V, as shown in Figure - 1 and 2. This law will not be followed through out the entire course of enzyme reaction, because as product accumulates (S_o - x) diminishes, and eventually will become comparable to Kn.

The general rate equation (5) integrated as
follows

$$Vt = \int \left\{ \frac{Kn}{(S_0 - x)} + 1 \right\} dx + constant \qquad (8)$$

When the constant is determined by the boundary condition
$x = 0$, $t = 0$, the solution is

$$Vt = Kn \ln \frac{S_0}{S_0 - x} + x \qquad (9)$$

In Figures - 1 and 2, it is clear that $v = 1.85$
I.U./liter/min. for 5 - nucleotidase from liver cirrhotic
sera and $v = 0.15$ I.U./liter/min. for normal enzyme at
$t = 10$ min., the amount of inorganic phosphate produced
in the reaction catalyzed by 5 - nucleotidase from liver
cirrhotic sera per min., is larger than of normal sera,
and the incubation of the enzyme with 1.0 mM A - 5 - MP
for 30 min., is in the region of zero order kinetics for
both systems.

Optimal A - 5 - MP concentration for 5 - nucelotidase

Figure - 3, shows that 1.0 nM A - 5 - MP is
optimum substrate concentration for both sources of
5 - nucleotidase (normal and liver cirrhotic) at pH
7.5, and 37°. The optimum substrate concentration
obtained in this studies (serum), is not yet reported
in the literature for both normal and liver cirrhotic.

Many standard assays in the literature uses this
optimum concentration in the reaction mixture catalyzed
by 5 - nucleotidase[99]. The optimum substrate concentration in general is recommended to obtaine more accurate
and consistent results for enzyme assays in serum[149].
Similarties in substrate optimum concentration for both
normal and liver cirrhotic 5 - nucleotidase in the
present studies, may not be due to a similarity in the
enzyme behavior, as at this specified optimum (S) there
was a difference in activity for both systems related
to the structural changes of the enzyme.

Metal requirements for 5 - nucleotidase

In the present work, Mg^{+2} and Mn^{+2} were found to
activate the enzyme 5 - nucleotidase in both normal and

liver cirrhotic sera, as shown in Table - 4, the degree
of activation by Mg^{+2} for liver cirrhotic sera was to
some extent larger (29.95%) than normal (23.25%).
Mn^{+2} activate 5 - nucleotidase to higher degree than Mg^{+2}.
Ni^{+2}, at 10 mM concentration completely inhibit the enzyme
in both normal and liver cirrhotic sera.

With regard to activation by Mg^{+2}, Mn^{+2} and inhibition by Ni^{+2}, snake venom 5 - nucleotidase was also found
to be activated by Mg^{+2} and Mn^{+2}, inhibited by Ni^{+2} and
Zn^{+2} [105]. 5 - nucleotidase has been studied in liver
and unlike other 5 - nucleotidases, divalent cations, as
Co^{+2}, Mn^{+2}, Mg^{+2} have no activating effect, these results
indicate enzyme specificity with regard to the source [105].

Levin and Bodansky[112] suggested that 5 - nucleotidase
is considered to have four sites relevant to its action
on nucleotides, one of these sites is for the combination
of Mg^{+2} with the enzyme, forming a bridge between the
enzyme and substrate.

The same mechanism may be applied to the action of
Mn^{+2} ions.

It may be possible to get direct information about the bridge formation in enzymes, by studying the binding of Ni^{+2} which inhibit completely the enzyme in both normal and liver cirrhotic sera.

The inhibition by Ni^{+2} could be used for estimation of 5 - nucleotidase activity, which depends on substracting the activity in the presence of Ni^{+2} from that in its absence, as Ni^{+2} has no effect on the activity of non - specific AP.

The activation by metal ions, has been studied very extensively by many workers. In 1942 Warburg and Christian[150] suggested that Mg^{+2} activation at constant enzyme and substrate concentrations, can be accounted for by the following equation

$$v = k\,(\,M\,)\,/\,(\,K_a\,+\,(\,M\,)\,) \qquad (\,10\,)$$

where k and K_a = constants

(M)   = concentration of activating metal ion.

This equation has been applied to a greater number of enzymes. Equation (10) by Warburg and Christian

considered to represents the ^{mass} action law for the
formation of an active complex between protein and metal
ion, this is true also when we consider the activation
of 5 – nucleotidase by Mg^{+2} or Mn^{+2}. 5 – nucleotidase
activation may be due to a reaction of Mg^{+2} with the
enzyme, with substrate, with the enzyme – substrate
complex, and this makes it possible to formulate a number
of alternative reaction mechanism one involve a complex
formation between enzyme (E) and metal ion (M), it
is further more assumed that the complex is the only
species catalytically active

$$E \ + \ M \ \underset{}{\overset{K_{(E)(M)}}{\rightleftharpoons}} \ EM \qquad\qquad (11)$$

$$EM \ + \ S \ \underset{k_2}{\overset{k_1}{\rightleftharpoons}} \ EMS \ \xrightarrow{k_3} \ EM \ + \ P \qquad (12)$$

The kinetic equation for mechanism one could be
formulated in equation (13).

$$v = k_3 (E)/(\ 1 \ + \ Km/(S) \ + \ Km \ / \ K_{(E)(M)} \ (M)(S) \ ) \qquad (13)$$

At constant (S) equation (13) can be written

$$v = \frac{k_3\,(E)(M)\,/\,(1+C)}{C/K_{(E)(M)}\,(1+C)+(M)} \qquad (14)$$

where $C = Km\,/\,(S)$

The second simple activation mechanism of 5 - nucleotidase by Mg^{+2} or Mn^{+2} involves the formation of Mg^{+2} - A - 5 - MP (Mn^{+2} - A - 5 - MP) complex which is assumed to be the true substrate reacting with the enzyme

$$S\ +\ M\ \underset{\longleftarrow}{\overset{K_{(S)(M)}}{\longrightarrow}}\ SM \qquad (15)$$

$$E\ +\ M\ \underset{k_2}{\overset{k_1}{\rightleftharpoons}}\ ESM\ \xrightarrow{\ k_3\ }\ E\ +\ PM \qquad (16)$$

This mechanism yields the following rate equation

$$v = k_3\,(E)\,/\,(\,1 + Km/(S) + Km\,/\,K_{(S)(M)}\,(M)(S)\,) \qquad (17)$$

Since equation (17) is kinetically completely identical with equation (13), this mechanism is also consistant with the emprical rate law. The only difference between the two equations is the constant for the metal - enzyme complex and the other that for metal - substrate complex.

A third simple mechanism — namely, combination of Mg^{+2} with the E — A — 5 — MP complex yields a similar form of rate equation and is therefore, kinetically difficult to distinguish from the other two mechanisms. The above mechanisms are postulated according to the various reports in literature, no experimental evidence are available to support these suggestion. Activation by metals could be used to differentiate between 5 — nucleotidase of different sources, liver cirrhotic and normal since they show different degree of activation.

Activation studies on 5 — nucleotidase

Investigation of the reaction rate of 5 — nucleotidase in presence of Mg^{+2} and varying concentrations of A — 5 — MP, Mg^{+2} (20 mM) was found to act as activater to the reaction at all substrate concentrations used, the mechanism by which Mn^{+2} ions activate the reaction rate is mentioned previously. The Km in the presence of Mg^{+2} is lower than in its absence (Figures 21, 22) in both systems, this indicate that in the presence of Mg^{+2} , the affinity of the substrate towards the enzyme is increased in both systems. The degree of activation was about two times higher for liver cirrhotic sera than normal (Table — 8).

1.0 mM Mn^{+2} was found to activate $\bar{5}$ - nucleotidase
of both systems (Figures 23, 24). Lower concentration
of Mn^{+2} than Mg^{+2} was required to activate $\bar{5}$ - nucleotidase. An intermediate complex of Mn^{+2} - A - 5 - MP
assumed to be formed at pH 7.5, and bond with the active
center of $\bar{5}$ - nucleotidase, and thus, permit the liberation of the phosphate group.

Km (A - 5 - MP) for $\bar{5}$ - nucleotidase in normal and
liver cirrhotic sera.

Very practical definition of Km is that, it is the
substrate concentration at half maximum velocity, this
is shown algebraically by

$$0.5 = \frac{1.0\,(S)}{Km + (S)}$$

$$0.5\,Km + 0.5\,(S) = 1.0\,(S)$$

$$Km = (S)$$

An obvious dependence of the velocity on the relative magnitude of Km or (S), if (S) is very large in
relation to Km, the expression becomes.

$$v = \frac{V(S)}{(S)}$$

and the velocity is maximum, independent on substrate concentration and therefore is zero – order, if the converse is true and Km is large the relation becomes

$$v = \frac{V(S)}{Km}$$

v depends on substrate concentration and the reaction is first order.

In summary the Km is by far the most fundamental constant used in enzyme chemistry.

Plotting the kinetic data of enzymatic reaction catalyzed by 5 – nucleotidase.

There are several ways of determining graphically the effects of varying (S) on the velocity of 5 – nucleotidase catalyzed reaction. The curves in Figures 4 – 7, based on various forms of Michaelis – Menten equation.

Many investigators have preferred to use the linear plots which are based on the linear transformation of the Michaelis – Menten equation, these plotts are applicable to the reactions catalyzed by 5 – nucleotidase a-   Lineweaver- Burk reciprocal plot, is based on the following linear transformation of the Michaelis - Menten equation

$$\frac{V}{v} = \frac{Kn + (S)}{(S)} \tag{18}$$

Cross - multiplying

$$\frac{1}{v} = \frac{Kn + (S)}{V(S)} \tag{19}$$

Separating terms

$$\frac{1}{v} = \frac{Kn}{V(S)} + \frac{(S)}{V(S)} \tag{20}$$

or

$$\frac{1}{v} = \frac{Kn}{V} \cdot \frac{1}{(S)} + \frac{1}{V} \tag{21}$$

This form is the equation for a straight line
($y = m\,x + b$). This if we plot y versus x
where $y = 1/v$ and $x = 1/(S)$, then m (the slope)
$= Kn/V$, and b (the intercept on the y axis)
$= 1/V$. We can also see that when $y = 0$ (i.e.,
$1/v = 0$), $mx = -b$ (i.e., Kn/V x $1/(S) = -1/V$)

and $x = -b/m$ (i.e., $1/(S) = - \dfrac{1}{V}$ x V/Km or
$1/(S) = - 1/Km$). This equation and the Figures
drown to show its application was used through this
study, such as Figure 7.

b— Woolf plot, which is based on the further mani-
pulation of the Lineweaver – Burk equation

$$\frac{1}{v} = \frac{Km}{V} \frac{1}{(S)} + \frac{1}{V}$$

Multiplying by (S)

$$\frac{(S)}{v} = \frac{(S) Km}{V} \frac{1}{(S)} + \frac{(S)}{V} \qquad (22)$$

$$\frac{(S)}{v} = \frac{Km}{V} + \frac{(S)}{V} = \frac{Km}{V} + \frac{1}{V}(S) \quad (23)$$

$$\frac{(S)}{v} = \frac{1}{V}(S) + \frac{Km}{V} \qquad (24)$$

the above equation has the general form for a straight line, $y = m x + b$, where
$$y = \frac{(S)}{v} \ , \quad m = \frac{1}{V}, \quad x = (S), \quad b = \frac{Km}{V}$$
where

$$\frac{(S)}{v} = 0, \ \frac{1}{V}(S) = \frac{-Km}{V} \ , \ \frac{(S)}{V} = \frac{-Km}{V} \ ,$$
$(S) = - Km.$

i.e., the intercept on the (S) axis gives - Km.

o- Hofstec plot is obtained as follows

$$v = \frac{V(S)}{Km + (S)}$$

Dividing numerator and denominator by (S).

$$v = \frac{V(S)/(S)}{(Km + (S)/(S)} = \frac{V}{(Km + (S))/(S)} \tag{25}$$

$$= \frac{V}{Km/(S) + 1} \tag{26}$$

Cross - multiplying

$$v \left[\frac{Km}{(S)} + 1 \right] = V \tag{27}$$

$$v = - Km \frac{v}{(S)} + V \tag{28}$$

in this form $y = v$, $m = - Km$, $x = v/(S)$, and $b = V$. We can see that when $v/(S) = 0$, $v = V$.

Similarly, when $v = 0$, $v/(S) = V/Km$, of these the double reciprocal plot where $\frac{1}{v}$ against $\frac{1}{(S)}$

has been by far the most widely used one. Eisenthal
and Cornish - Bowden[139] in 1974 have introduced a
new plot for analyzing the result of kinetic experiment in which the Michaelis - Menten equation is
obeyed. This method depends on plotting of observations ((S) and (V)) as lines in parameter space,
instead of points in observation space (Figure 6);
it has many advantages over the traditional methods
of plotting kinetics results. It is very simple
to use and require no mathematical calculations;
Moreover, it provides a clear and accurate information about the accuracy of the experiment. It also
provides an indication of the kinetic constants
and finally, when used carefully, the kinetic estimates made by this method, are the same as these
provided by computer programme[139].

In the present work the graphical procedure was
compared with three other linear plotts to determine
the Km (A - 5 - MP) of $\bar{5}$ - nucleotidase in normal
or cirrhotic sera.

The Km (A - 5 - MP) of normal 5 - nucleotidase

It was found to be equal to 0.18 + 0.03 mM (Table - 5), this value is similar to that obtained by Bergmeyer[15?] who reported a Km value of 0.2 mM in serum.   Other Km in different sources were obtained, the Km values for yeast 5 - nucleotidase[116] are < 0.2 mM for A - 5 - MP, the increased or decreased Km (A - 5 - MP) values obtained in the present studies compared to that obtained by other workers using different sources suggest the presence of different enzyme distributed in different biological systems due to various affinities of these enzymes to their substrates, an increase Km value for example indicate a lower affinity of the enzyme to substrate, a decrease Km value indicates a high affinity of the enzyme to substrate.

Determination of Km (A - 5 - MP) of 5 - nucleotidase in liver cirrhotic

Table - 5 shows that Km (A - 5 - MP) is 0.06 + 0.02 which is greatly reduced from Km (A - 5 - MP) in normal constant.   This indicates that 5 - nucleotidase in liver cirrhotic has a higher affinity for A - 5 - MP than the corresponding enzyme in normal individuals, again the

difference in the affinity of enzyme to substrate suggest
the presence of different enzyme with different structural
make up.

Activity of 5 – nucleotidase towards 2 – d – A – 5 – MP in normal and liver cirrhotic sera.

In an attempt to study the affinity of 5 – nucleotidase towards compounds other than A – 5 – MP, 2–d–A–5–MP
was used as substrate for the reaction. Table (6A)
reveals that 5 – nucleotidase activity of liver cirrhotic
persons was much higher than that of normal individuals,
and the activities ratios towards 2 – d – A – 5 – MP, for
liver cirrhotic to normal were larger than one, Kn
(2 – d – A – 5 – MP) of normal and liver cirrhotic
5 – nucleotidase was found to be 0.35 nM, 0.25 nM respectively, (Table – 6B). This show that Kn (2–d–A–5–MP)
is markedly reduced for liver cirrhotic sera, indicating
higher affinity for 2 – d – A – 5 – MP than corresponding
enzyme in normal person. This may be again due to a
difference in the structural organization of the enzyme
or else, it may be attributed to a difference in isoenzyme composition of 5 – nucleotidase in normal and liver
cirrhotic individuals, no studies has been done on the
isoenzyme distribution of 5 – nucleotidase in serum.

The existance of two types of $\bar{5}$ - nucleotidase in these systems with markedly different Km values is suggested from present studies, the one with low Km located in the serum of liver cirrhotic sera and the other with higher Km in the serum of normal individuals.

Effect of serum concentration

At any constant substrate concentration, the rate of the reaction increases linearily with increasing enzyme concentration (Figure - 9), that is, the rate of enzyme - catalyzed reaction conversion of substrate to product is (under assay conditions) directly proportional to the enzyme concentration, (v = ke). This linear relationship between initial velocity and enzyme concentration will in general be observed when the following conditions are met.

1.· There is no inhibitor or toxic impurity present in the reaction mixture that would inactivate some of the enzyme present.

2— The concentrations of substrate and any required cofactor are greatly in excess of the enzyme concentration.

3-   The method being employed to follow the reaction
      rate truly reflects the velocity of substrate — to-
      product conversion at all enzyme concentrations used.

Inhibition of $\bar{5}$ - nucleotidase

Classically the study of inhibitory effects on
isolated enzymatic reactions and on metabolic sequences
in general has been of the greatest importance in establi-
shing the nature of the free reactants, the nature of
their binding sites on the enzyme, the specificity and
mechanism of the reaction. In the cell, inhibition of
key reactions by substances that may be product of reaction itself,   provide a ready and delicate poised control
mechanism for the maintenance of a relatively constant
intracellular environment.

The kinetics of enzymic reactions catalyzed by
$\bar{5}$ - nucleotidase in the presence of varying concentrations
of different inhibitors (Ni^{+2}, ATP, adenosine) were
investigated in this work.

Inhibition by Nickel chloride

The effect of various concentrations of nickel chloride was investigated at pH 7.5. It was found that Ni^{+2} ions inhibits 5 - nucleotidase to varying extent according to its concentration. The inhibition being much more pronounced on 5 - nucleotidase for pathological system. The type of inhibition was also investigated and determined according to Dixon plot; (plot of 1/v versus concentration of inhibitor), since under the conditions of the experiments it was desirable to vary the inhibitor concentrations and to maintain a constant substrate concentration, the initial velocities were determined for two substrates concentrations (and varying inhibitor concentrations), Ki was directly determined. The inhibition for both systems was of competitive nature and Ki from table - 7 for normal was lower than that of liver cirrhotic sera.

The competitive inhibitor is a compound that combines with the enzyme in such a way as to prevent the enzyme from combining with the substrate. Frequently competitive inhibitors are either non - metabolizable or they are alternate substrates of the enzyme. Because such inhibitors resemble the " true " substrate structurally, they compete with the substrate for the active site.

Ni^{+2} ions bear no structural relationship to
$A - 5 - MP$, yet they yield competitive inhibition kinetics.
There are many compounds that yield competitive type, yet
they bear no structural relationship to the substrate such
as end - product, or near end - product of a metabolitic
pathway; metals..... etc. The combination of the Ni^{+2}
with $\bar{5}$ - nucleotidase causes a change in the " conformation "
tertiary or quaternary structure of the enzyme, that results
in a decrease affinity for $A - 5 - MP$.

$\bar{5}$ - Nucleotidase at any time is present in three
forms, free ($\bar{5}$ - Nucleotidase), $\bar{5}$ - Nucleotidase -
$A - 5 - MP$ complex; and $\bar{5}$ - Nucleotidase - Ni^{+2} complex.

$$(\bar{5} - \text{Nucleotidase})_{T} = (\bar{5} - \text{Nucleotidase}) \cdot$$
$$+ (\bar{5}-\text{Nucleotidase-A5-MP}) + (\bar{5} - \text{Nucleotidase} - Ni^{+2})$$

$$\bar{5} - \text{Nucleotidase} = (\bar{5} - \text{Nucleotidase})_{T} - (\bar{5}-\text{Nucleotidase} - Ni^{+2})$$
$$- (\bar{5} - \text{Nucleotidase} - A - 5\ MP)$$

Dividing by ($\bar{5}$ - Nucleotidase - $A - 5 - MP$)

If $\bar{5} - N$ = $\bar{5}$ - Nucleotidase

$$\frac{(\bar{5}-N)}{(\bar{5}-N-A-5-MP)} = \frac{(\bar{5}-N)_T - (\bar{5}-N-A-5-MP) - (\bar{5}-N-Ni^{+2})}{(\bar{5}-N-A-5-MP)}$$

$$(29)$$

$$\frac{\bar{5}-N}{(\bar{5}-N-A-5-MP)} = \frac{Kn}{A-5-MP}$$

$$(29)$$

$$\therefore \frac{Kn}{A-5-MP} = \frac{(\bar{5}-N)_T}{(\bar{5}-N-A-5-MP)} - \frac{(\bar{5}-N-A-5-MP)}{(\bar{5}-N-A-5-MP)}$$

$$- \frac{(\bar{5}-N-Ni^{+2})}{(\bar{5}-N-A-5-MP)}$$

$$(30)$$

$$(\bar{5}-N-Ni^{+2}) = \frac{(\bar{5}-N)(Ni^{+2})}{Ki}$$

$$(40)$$

$$\frac{1}{(\bar{5}-N-A-5-MP)} = \frac{Kn}{(\bar{5}-N)(A-5-MP)}$$

$$(41)$$

$$\therefore \frac{Kn}{(A-5-MP)} = \frac{(\bar{5}-N)_T}{(\bar{5}-N-A-5-MP)} - 1 - \frac{(\bar{5}-N)(Ni^{+2})}{Ki}$$

$$x \frac{Kn}{(\bar{5}-N)(A-5-MP)}$$

$$(42)$$

$$\frac{(\bar{5}-N)_T}{(\bar{5}-N-A-5-MP)} = \frac{Kn}{(A-5-MP)} + 1 + \frac{Kn\,(Ni^{+2})\,(\bar{5}-N)}{(A-5-MP)\,(\bar{5}-N)\,Ki}$$

$$(43)$$

Substituting V/v for $(\bar{5}-N)_T / (\bar{5}-N-A-5-MP)$

$$\frac{V}{v} = \frac{Kn}{(A-5-MP)} + \frac{Kn\,(Ni^{+2})}{(A-5-MP)\,Ki} + 1 \qquad (44)$$

Multiplying both sides of the equation by $(A-5-MP)$

$$\frac{V\,(A-5-MP)}{v} = \frac{Kn\,(A-5-MP)}{(A-5-MP)} + \frac{Kn\,(Ni^{+2})\,(A-5-MP)}{Ki\,(A-5-MP)} + A - 5 -$$

$$(45)$$

Cancilling $(A-5-MP)$ terms and factoring out Kn

$$\frac{V\,(A-5-MP)}{v} = Kn\left(1 + \frac{(Ni^{+2})}{Ki}\right) + A - 5 - MP \qquad (46)$$

Inverting

$$\frac{v}{V\,(A-5-MP)} = \frac{1}{Kn\left(1 + \frac{(Ni^{+2})}{Ki}\right) + (A-5-MP)} \qquad (47)$$

In the above equation, Ki represents the dissociation
constant for the $\bar{5}$ - nucleotidase - Ni^{+2} complex. The
lower the value of Ki in the normal, the greater is the
affinity of the Ni^{+2} for $\bar{5}$ - nucleotidase.

Inhibition by ATP in both normal and liver cirrhotic sera

ATP, is a biological derivative of phosphorus plays
a key role in the energy transactions of living organisms,
it is the basic unit of energy. In our investigations
ATP was found to be a competitive inhibitor of $\bar{5}$ - nucleotidase, (Figures 13, 14) which compete the substrate
(A - 5 - MP) for the active site. The inhibitor constants
were found to be 0.1 + 0.03 for normal and 0.225 + 0.2 for
liver cirrhotic; i.e., about 2 times higher than normal.
The degree of inhibition was found to be greater for normal
than liver cirrhotic (Figure - 15).

The equation for the competitive case which is applied
to ATP inhibition is

$$\frac{1}{v} = \frac{Km}{V(S)} + \frac{1}{V} + \frac{Km}{V(S)} \quad \frac{(i)}{Ki} \qquad (48)$$

which gives a straight line on plotting 1/v against (i).

The difference in Ki values confirm the earlier suggestion of the presence of different enzyme system with hew characteristic, which differs from the normal.

Inhibition by adenosine (End – product inhibitor)

Inhibition of $\bar{5}$ – nucleotidase by adenosine is of non – competitive type, this is very well illustrated in (Figures 17, 18) for both normal and liver cirrhotic sera.

The rate equation of the non – competitive type which may be applied to our system is the following

$$v = \frac{V \, (\, S \,)}{(\, Km \, + \, (S) \,) \, (\, 1 \, + \, \frac{(i)}{Ki} \,)} \qquad (\, 49 \,)$$

This becomes

$$\frac{dx}{dt} = \frac{V \, (\, S_o \, - \, x \,)}{(\, Km \, + \, S_o \, - \, x \,) \, (\, 1 \, + \, \frac{x}{Ki} \,)} \qquad (\, 50 \,)$$

Dixon has given simple graphical method which gives Ki directly without calculation which is illustrated in (Figures 17, 18), when the velocity was determined with a series of adenosine concentration, keeping the substrate concentration constant (A - 5 - MP), a straight line is obtained in plotting $\frac{1}{v}$ against (i), the intersection of the lines gives - Ki which lies on the base line, this can be seen by putting $\frac{1}{v} = 0$ in the reciprocal form of the following equation

$$v = \frac{\dfrac{Ke}{1 + \dfrac{(i)}{Ki}}}{1 + \dfrac{Km}{(S)}} \qquad (51)$$

Gives

$$\frac{1}{v} = \frac{1}{V} \left(1 + \frac{Km}{(S)} \right) \left(1 + \frac{(i)}{Ki} \right) \qquad (52)$$

which represents the straight line of (Figures 17, 18).

It may be however to obtain Ki from a number of measurements in which both substrate and inhibitor concentrations varies according to Hunter and Downs[152] which enables all the data to be recalculated, so that they are brought on to a single line.

For each observation a velocity ratio α is defined by

$$\alpha = \frac{v_i}{v}$$

v_i = the velocity in the presence of inhibitor

v = the velocity in the absence of inhibitor

For the non-competitive case which is resulted from adenosine effect equation (51) gives

$$\frac{1 - \alpha}{1 - \alpha} = Ki$$

which is independent of substrate concentration. For normal individuals α can be calculated

$$\alpha = \frac{2.5}{6.5} = 0.384 \qquad \text{at 0.8 nM adenosine}$$

For liver cirrhotic patients

$$\alpha = \frac{65}{90} = 0.722 \qquad \text{at 0.8 nM adenosine}$$

The non-competitive inhibitor (adenosine) can combine with either the free $\bar{5}$ - nucleotidase or with $\bar{5}$ - nucleotidase - A - 5 - MP complex. The adenosine combines with the enzyme somewhere other than at the

active site. In the presence of adenosine as non - competitive inhibitor, the velocity still depends on (A - 5   MP)
until saturation is attained. However, the degree of inhibition at any adenosine concentration is independent of
(A - 5 - MP) and depends only on (Adenosine) and Ki.
The Km is unchanged , V , however, is decreased

$$\frac{v_i}{V} = \frac{Ki}{Ki + (\text{ Adenosine })} \qquad (53)$$

Such differences in inhibitory characteristics, with
nickel chloride, ATP and adenosine in both normal and liver
cirrhotic sera, worth further investigations, for it may
prove to be of great diagnostic significance.

Effect of temperature on 5 - nucleotidase activity

It was found that, as the temperature increased the
rate of the reaction catalyzed by 5 - nucleotidase in both
systems increased till $37^{\circ} - 40^{\circ}$ then it started to decrease and reached zero at 100° (Figure 25) due to enzyme
denaturation.

There was a difference in the velocity at different
temperatures between the normal and liver cirrhotic sera,

this result may be attributed to the presence of higher quantity of 5 – nucleotidase in liver cirrhotic sera, or due to the presence of activator in liver cirrhotic sera which results in increasing the activity of this enzyme.

The kinetic effect of temperature on this enzyme was passed through 3 stages, because enzyme denaturation generally leads to a decrease in activity while the kinetic effect produces an increase in activity with increasing temperature, it is evident that for any time of exposure to given temperature, the rate of an enzymic reaction might appear to increase at first, go through a maximum, and eventually decline for both systems (Figure 25).

The magnitude of temperature effect is frequently reported in terms of temperature coefficient, which may be defined as the ratio of the reaction velocity at $t + 10^{o}$ to that t^{o}, and denoted by Q_{10}, thus

$$Q_{10} = \frac{K_{t+10}}{K_t} \qquad\qquad (54)$$

where K_{t+10} and K_t are the rate constants at respective temperatures $t + 10$ and t^{o}. The Q_{10} values does vary with temperature, the range of temperatures for which the measurement was made should always be quoted.

The Arrhenius equation

The plot of log V versus 1/T gave a straight line as shown in (Figure - 26) which is the Arrhenius plot. At high (S), v becomes V and equal to $k_3 e$, and k_3 should obey the Arrhenius equation

$$\ln k = \frac{-E}{RT} + \text{Constant} \qquad (55)$$

The above equation is based on Arrhenius assumption that when reaction occurs an equilibrium exists between the reactants and an intermediate collision complex, or activated complex, which has an energy content higher than that of the reactants.

pH Studies on 5 - nucleotidase in normal and liver cirrhotic individuals.

When both normal and liver cirrhotic sera were exposed to different pH values, and the substrate - velocity relationships of 5 - nucleotidase were determined at each pH value (in the range 7 - 8.5[*]), it was observed that

---

[*] It was decided to use this range of pH because of the precipitation of nickel, probably as phosphate and hydroxide, at pH higher than 8.5.[153]

both serum $\bar{5}$ - nucleotidase exhibited pH dependent characteristics; in that there was an optimum substrate concentration (A - 5 - MP) at each pH value (Figure - 27),
these optima shifted towards higher concentrations as
the pH is increased from 7 - 8.5, this shift of optimum
(S) towards higher values with increasing pH, indicate
a lower affinity of the enzyme for its substrate (A-5-MP),
so that the substrate concentration needed to half-saturate
the enzyme increases with increasing pH values subsequently
Km values are increased. The shift in optimum (S) with
pH however, was found to be sharper in normal sera than
liver cirrhotic sera (Figure - 28).

The pK of the groups on $\bar{5}$ - nucleotidase present
in or near the active center could be determined by plotting log v versus pH (Figure - 29), from this figure
the pK values for liver cirrhotic $\bar{5}$ - nucleotidase was
found to be 7.4 and 7.2 for normal $\bar{5}$ - nucleotidase. The
small shift in the pK value indicates a difference in the
environment in the vicinity of the group undergoing ionization, these pK values may very well be due to the resi-
dues present at the active sites, and it may be related
to the histidine, or α - ammonium group, which their pK
values present in the range.

Plots of pKm versus pH (Figure - 30) gives the
same indication, and show great dependency of Km on pH
in the pH range 7 - 8.5 for both types of serum $\bar{5}$ - nucleotidase. According to the difference between the two curves.
This difference in the curves confirm the earlier suggestion of the presence of different enzymes in both systems.

pH Optimum determination in the presence and absence of
20 mM Mg^{+2} ions at 37°

Figures 31, 32, show that, the velocities at the
respected pH optimum were higher in the presence of Mg^{+2}
ions than in its absence, for both normal and liver cirrhotic sera, This difference in activity value confirm the
earlier suggestion of the bridge mechanism between Mg^{+2} -
A - 5 - MP and the $\bar{5}$ - nucleotidase, and thus increasing
the affinity of the substrate towards the enzyme.

$\bar{5}$ - Nucleotidase is extremely sensitive to pH; its
activity is diminished at either side of a relatively narrow
range (Figures 31, 32). These effects are due to combination of three factors:

1-   Irreversible effects of extremes pH on protein
       structure, including alterations of the strength
       and mode of binding of $\bar{5}$ - nucleotidase.

2-   Effect on the ionization of the substrate (A-5-MP)

3-   Effects on its binding to the enzyme and on reacti-
       vity in catalysis.

The initial velocity of the enzymatic reaction catalyzed by $\overline{5}$ - nucleotidase exhibits these distinct phases
as a function of pH (Figures 31, 32.), a region of pH
(at low values) where there is an increase, a region
(at high values) where there is a decrease, and an
intermediate range (usually neutrality) when the activity
is maximal and approximately constant, the pH optimum
(7.5 - 8) leading to a characteristic bell - shaped
curve. The same range of pH optimum was reported by
Campbell[99] and Young[97], which supports our results.

Determination of interaction coefficient n,   between ATP
binding sites, for both normal and liver cirrhotic patients.

From Hill plot[141] of coordinate a straight line
of positive slope n = 1 was obtained (Figure - 33),
for normal $\overline{5}$ - nucleotidase, in the absence of ATP, but
in the presence of ATP, Hill plot of coordinate gave a
straight line with interaction coefficient equals to 1.473
for normal, which suggest the cooperative effect between

ATP molecules, but the interaction coefficient remained equals to one for liver cirrhotic sera. This difference between the two values of interaction coefficient for both systems in the presence of ATP, suggest the presence of different enzyme in normal and liver cirrhotic sera.

CHAPTER FIVE

S U M M A R Y

S U M M A R Y

1—   Assay of serum $\bar{5}$ – nucleotidase in normal and liver cirrhotic sera.

$\bar{5}$ – Nucleotidase activity was elevated in patients with liver cirrhosis, greater values of $\bar{5}$ – Nucleotidase activity were found in biliary cirrhosis.

2— $\bar{5}$ – Nucleotidase kinetics at 37° in normal and liver cirrhotic patients.

   a. $\bar{5}$ – Nucleotidase from liver cirrhotic sera was less stable than the normal enzyme.

   b. The velocity of $\bar{5}$ – nucleotidase for liver cirrhotic sera per minute, at $t = 10$, was greater than normal controls.

   c. The optimum (S) for $\bar{5}$ – nucleotidase was found to be 1.0 mM A – 5 – MP, for both normal and liver cirrhotic sera.

   d. The substrate – velocity relationship of $\bar{5}$ – nucleotidase was found to be characteristic of the enzymes which obey Michaelis – Menten kinetics, for both systems.

e. At any (A – 5 – MP) concentration, patients with liver cirrhosis, exhibited higher reaction rates with respect to $\bar{5}$ – nucleotidase than normal.

f. Mg^{+2}, Mn^{+2} were found to activate both normal and liver cirrhotic $\bar{5}$ – nucleotidase but Ni^{+2}, inhibited the enzyme in both systems.

g. Km (A – 5 – MP) of $\bar{5}$ – nucleotidase was found to be significantly lower in patients with liver cirrhosis, than $\bar{5}$ – nucleotidase of normal controls.

h. Km ($\bar{2}$ – d – A – 5 – MP) of $\bar{5}$ – nucleotidase was found to be lower in patients with liver cirrhosis, than normal controls.

i. At any $\bar{5}$ – nucleotidase concentration, patients with liver cirrhosis, exhibited higher reaction rates than normal persons.

j. Both ATP and adenosine act as inhibitors to $\bar{5}$ – nucleotidase. The inhibitory constants for ATP and adenosine were different in normal and liver cirrhotic individuals.

k. There was a difference in the velocity of $\bar{5}$ – nucleotidase reaction, at different temperatures between normal and liver cirrhotic sera.

1. The substrate – velocity relationship of $\overline{5}$ – nucleotidase, was found to be very much pH dependent
   for both normal and liver cirrhotic sera.

n. The nature of the enzyme was changed in the
   presence of ATP for normal $\overline{5}$ – nucleotidase only.

CHAPTER SIX

REFERENCES

REFERENCES

1. Thorpe, W.V., Bray, H.G. James, (1970), In Bio-chemistry for Medical students, p. 158, Churchil, London.

2. Green-berg, D.M., Harper, A.H., (1960), In Enzymes in Health and Disease, p. 26, Charles C Thomas, U. S. A.

3. Pauling, L., Itano, H.A., Singer, S.J., and Wells, I.G., (1949), Science, 110, 543 - 548.

4. Hoffman, W.S., (1970), In the Biochemistry of Clinical Medicine. 4th edition p. 182, 462, Year Book, Chicago.

5. Physick John Syng. Reported in S. Solis - Cohen, and Githens. T.S., (1928), Pharmaco - therapeutics D.A. Appleton Co., New York.

6. Reiser, H.G., Patton, R. and Roetting, L.C., (1951), A. M. A. Arch. Surg., 63 - 568 - 575.

7. Green-berg, D.M., Harper, A.H., (1960), In Enzymes in Health and Disease, p. 213, Charles C Thomas, U. S. A.

8. Raab, W., (1935), Klin, Wehnschr, 14, 1633 -- 1635.

9. Raab, W., and Breuer, J., (1935), Ztschr. fid. Ges. Exper. Med., 97, 415 -- 422.

10. Gurchot, C., Krebs, E.T., Jr., and Krebs, E.T., (1947), Surg., Gynec. and obst., 84, 301 - 312.

11. Krebs, E.T. and Krebs, E. T., Jr., and Gurchot, C., (1947), Med. Rec., 160, 479 - 480.

12. Mcallister R.A., (1970), In Enzymes and the Determination of Enzyme Activity, pp. 1 - 2, Butherwarth, London.

13. Green-berg, D.M., Harper, A.H., (1960), In Enzymes in Health and Disease, p. 344, Charles C Thomas, U.S.A.

14. Zierler, K.L., (1958), Bull. Johns Hopkins Hosp., 102, pp. 17 - 20.

15. Zierler, K.L., (1958), Ann. New York Acad. Sci.,
     75, pp. 227 - 234.

16. Schlamowitz, M., (1958), Ann. New York Acad. Sci.,
     75, pp. 373 - 379.

17. Vessell, E.S. and Bearn, A.G., (1958), Ann. New
     York Acad. Sci., 75, 286 - 291.

18. Strandjord, P.E., Thomas, K.E., et al., (1960),
     In Serum Enzymes VI. The release and activation
     of LDH and ICD from myocardial infarction.

19. Green-berg, D.M., Harper, A.H., (1960), In Enzymes
     in Health and Disease p. 348, Charles C Thomas, U.S.A.

20. Bergmeyer, H.U., (1974), In Methods of Enzymatic
     Analysis, vol. I, pp. 6 - 7, Academic Press,
     New York.

21. Born, H.H., (1968), In Enzymes in Urine and Kidney,
     p. 119.

22. Lee, D.A., Cockett, A.T.K., Caplan, B.M., and Chianori, N., (1966), J. Urol., 95, 77.

23. Searcy, R.L., (1969), In Diagnostic Biochemistry, p. 510, McGraw - Hill Co., New York.

24. Searcy, R.L., (1969), In Diagnostic Biochemistry, p. 17, McGraw -- Hill Co., New York.

25. Smyrniotis, F., Schenker, S., O'Donnell, J., and Schiff, L., (1962), Amer., J., Dig. Dis., 7, 712.

26. Searcy, R.L., (1969), In Diagnostic Biochemistry, p. 344, McGraw - Hill Co., New York.

27. Abbasy, A.S., and Aboulwafa, M.H., (1961), J. Pediat., 59, 60.

28. Searcy, R.L., (1969), In Diagnostic Biochemistry, p. 336, McGraw - Hill Co., New York.

29. Thorpe, W.V., Bray, H.G., James, S.P., (1970), In Biochemistry for Medical students, p. 245. Churchil, London.

30. DeRitis, F., et al., (1958), Lancet, 11, 214.

31. Zimmerman, H.J., and West, M., (1963), Amer. J. Gastroent., 40, 387.

32. Searcy, R.L., (1969), In Diagnostic Biochemistry p. 41, MacGraw - Hill Co., New York.

33. Kowlessar, O.D., et al., (1961), Am. J. Med., 31, 231.

34. Schiff, L., M.D., (1975), In Diseases of the Liver, p. 227.

35. Albert, Z., et al., (1961), Nature, 191, 767.

36. Zeim, M., and Discombe, G., (1970), Lancet, 11, 748.

37. Szczeklik, E., et al., (1961), Gastroenterology, 4, 353.

38. Karmen, A., Wroblewsky, F., and Laduc, J.S., (1955), J. Clin Investigation, 34, pp. 126 - 133.

39.   Wroblewski, F., (1957), Amer. J. Mod. Sci.,
      234, 301.

40.   Brahn, B., (1916), Sitzher, Kgl. Preusse Akad,
      Wiss., 20, 478, (Chem. Abstr., 10 : 1762, 1916)

41.   Utmann, J.E., Hyman, G.A., Harvey, J.L., and Dente,
      A.R., (1957), Blood, 12, pp. 1114 -- 1121.

42.   Nachman sohn, D. ., and Wilson, I.B., (1955), In
      Molecular basis for generation of Bioelectric Poten-
      tials in electrochem. In Biology and Medicine,
      p. 167 -- 186, New York.

43.   McEwen, W.K., Kimura, S.J., and Feeney, M.L., (1958),
      Am. J. Ophth., 45, 67.

44.   Pirie, A., (1956), Brit. M. Bull., 12, 32.

45.   Morton, R.A., and Goodwin, T.W., (1956), Brit. M.
      Bull., 12, 37.

46.   Bohinski, R.C., (1973), In Modern Concepts in
      Biochemistry, p. 280, Allyn and Bacon, Inc., Boston.

47. Inner Field, I. Angrist, A., and Schwartz, A., (1953), J. A. M. A., 152, pp. 597 - 605.

48. Hopen, J.M., and Campagna, F.N., (1955), Am. J. Ophth., 40, pp. 209 - 214.

49. Hopen, J. M., (1954), Am. J. Ophthal., 38, pp. 84 - 87.

50. Greenberg, D.M., Harper, A.H., (1960), In Enzymes in Health and Disease, p. 219, Charles C Thomas, U. S. A.

51. Greenberg, D.M., Harper, A.H., (1960), In Enzymes in Health and Disease, p. 239, Charles C Thomas, U. S. A.

52. Bergmeyer, H.U., (1974), In Methods of Enzymatic Analysis, vol. I, p. 94, Academic Press, New York.

53. Enzyme Nomenclature, (1965), Recommendations of the International Union of Biochemistry on the Nomenclature and Classification of Enzyme, together with their Units and Symbols of Enzyme Kinetics, p. 6 - 14

54.   Dixon, M., and Webb, E.C., (1971), In the Enzymes, Chap. IV., Longman, London.

55.   Florkin, M., and Stotz, E.H. (1964), In Comprehensive Biochemistry vol. 12.   pp.   91 - 124.

56.   Zeffren, E. and Hall, P.L., (1973), In the Study of Enzyme Mechanisms, pp.   62 - 86.

57.   Bull, A.T., et al., (1974), In Companion to Biochemistry, pp. 227 - 247, Longman, London.

58.   Kohn, T. and J.L. Reis, (1963), J.   Bacteriology, 86, 713.

59.   Nou, H.C., and L.A. Hoppel, (1964), B B R C, 17, 215.

60.   Repaska, R., (1958), Biochim, Biophys.   Acta., 30, 225.

61.   Neu, H.C., (1967), J.   Biol.   Chem., 242, pp. 3905 - 3911.

62.   Olson, A.C., and H.J., Fraser, (1974), Biochim. Biophys.   Acta., 334, pp.   156 -- 167.

63. Rabin, E.L., Tenenhouse, H., and Fraser, M.J., (1972),
Biochim, Biophys. Acta., 259, pp. 50 - 68.

64. Björk, W., (1961), Biochim. Biophys. Acta.,
49, 195.

65. Sulkowski, E., W. Björk, and M. Laskowski, (1963),
J. Biol. Chem., 238, 2477.

66. Mann, T., (1945), Bioch. J., 39, 451.

67. Heppel, L.A., and R.J. Hilmoe, (1951), J. Biol.
Chem., 188, 665.

68. Bodansky, O., and M. K. Schwartz, (1963), J. Biol.
Chem., 238, 3420.

69. Pilcher, C.W., and T.G. Scott, (1967), Bioch. J.,
104, 41c.

70. Nori-Kaff, A.B., et al., (1950), Federation Proc.
9, 210.

71. Delamirande, G., et al., (1952), J. B. B. Cyto.,
4, 373.

72. Arsenis, C., and O. Touster, (1968), J. Biol. Chem., 243, 5702.

73. Itoh, R., A. Mitsui, and K. Tsushima, (1967), Biochim. Biophys. Acta., 146, 151.

74. Itoh, R., A. Mitsui, & K. Tsushima, (1968), J. B. (Tokyo), 63, 165.

75. Fritzon, P. (1967), Eur. J. B., 1, 12.

76. Fritzon, P. (1968), Biochim. Biophys. Acta., 151, 716.

77. Fritzon, P. (1969), Biochim. Biophys. Acta., 170, 534.

78. Segal, H.L., and Brenner, B.M., (1960), J. Biol. Chem., 235, 471.

79. Widnell, C.C., and J.C. Unkeless, (1968), Proc. Natl . Acad. Sci., 61, 1050.

80. Center, M.S., and F.J. Behal, (1966), A B B, 114, 414.

81. Burger, R.M., and J.M.Lowenstein, (1970), J. Biol.
     Chem., 245, 6274.

82. Reis, J., (1935), Bull. Soc. Chim. Biol., 16,
     385.

83. Reis, J.L., (1951), B. J., 48, 548.

84. Ipata, P.L., (1967), Nature, 214, 618.

85. Ipata, P.L., (1967), B B R C , 27, 337.

86. Ipata, P.L., (1968), Biochemistry, 7, 507.

87. Berne, R.M., (1963), Am. J. Physiol. 204, 317.

88. Rostgaard, J., and O. Behnke, (1965), J. Ultrastruc.
     Res., 12, 579.

89. Williamson, J.R., and D.L. Dipietro, (1965),
     B. J., 95, 226.

90. Baer, H.P., and G.I. Drummond, (1968), Proc. Soc.
     Exptl. Biol. Med. 127, 33.

91. Lisowski, J., (1964), Arch. Immunol. Therap.
     Exptl, 12, 542.

92. Lisowski, J., (1966), Arch. Immunol. Therap.
     Exptl., 14, 195.

93. Cozzani, I., P.L. Ipata, and M. Ranievi, (1969),
     F E B S Letters, 2, 189.

94. Hardonk, M.J., and J. Koudstaall, (1968), Histo-
     chemie, 15, 290.

95. Hardonk, M.J., and M.G.A. deBoer, (1968), Histo-
     chemie, 12, 29.

96. Theodore, F.D., and Mary Purdon, (1954), J. Clin.
     Path., 7, 341.

97. Young, I.I., (1958), Ann. New York Acad. Sci.,
     75, 357.

98. Bardawill, C., and Chang, C., (1964), Cand. Med.
     Assn. J., 89, 755.

99.   Campbell, D.M., (1962), Bioch. J., 82, 34p.

100. Neu, Harold, C. (1968), Biochemistry, 7, (10),
      3766.

101. Magnusson, Goran, (1971), Eur. J. Biochem., 20
      (2), 225.

102. Gercignani, Gioranni; Serra, (1974), Biochemistry,
      13 (17), 3628.

103. Lisowski (1966), Biochim. Biophys. Acta, 113,
      321.

104. Gulland, J.M., and Jackson, E.M., (1938), Bioch.
      J. 32, 597.

105. Boyer, P.D., et al., (1971), In the Enzymes,
      vol. IV, pp. 337 - 352.

106. Reis, J.L., (1937), Enzymologia, 2, 110.

107. Reis, J.L. (1950), B. J., 46, XXI

108. Kay, M.A.G. (1955), Biochim. Biophys. Acta, 18, 456.

109. Ahmed, Z., and J.L. Reis, (1958), B. J. 69, 386.

110. Hardonk, M.J., (1968), Histochemie, 12, 1.

111. Song, C.S., and O. Bodansky, (1967), J. Biol. Chem., 242, 694.

112. Levin, S.J., and O. Bodansky, (1966), J. Biol. Chem., 241, 51.

113. Dvorak, H.F., Y. Anraku, and L.A. Heppel, (1966), B B R C, 24, 628.

114. Dvorak, H.F., and L.A. Heppel, (1968), J. Biol. Chem., 243, 2647.

115. Murray, A.W., and B. Friedrichs, (1969), Bioch. J., 111, 83.

116. Takei, S., (1967), Agr. Biol. Chem., 310 (11), 1251.

117. Kluge, H., Hartmann, W., Wieczorck, V., (1972),
     J. Neurochem., 19 (5), 1409.

118. Ipata, P.L., (1967), Analyt. Biochem. 20, 30.

119. Goldberg, D.M., and G. Ellis, (1972), J. Clin.
     Path., 25, 907.

120. Persijn, J.P., Van der Slike, W., (1971), Int.
     Congr. Clin. Chem., 2, 108.

121. Kucerova, Z., Skoda, J. (1971), Biochim. Biophys.
     Acta, 247 (2), 194.

123. Secchi, G.C., A. Rezzonico, and N. Gervasini, (1967),
     Enzymol. Biol. Clin. 8 (1), 42.

124. Sherlock, S., (1968), In Diseases of the Liver
     and Biliary System, p. 398 - 401, 4th Edition.

125. Rubin, E., and Pappor, H., (1967), Medicine, 46,
     163.

126. Robbins, S.L., (1968), In Pathology, p. 912 - 943,
      3rd Edition, W.B. Saunders Company, London.

127. Whitby, and Britton, (1963), In Disorder of the
      Blood.

128. Himsworth, H.P., (1950), In the Liver and its
      Diseases, 2nd Edition.

129. Boyol, W., (1961), In Text Book of Pathology,
      p. 793, Seventh Edition, Henry Kimpton, London.

130. Sherlock, S., (1951), Brit. Heart, J., 13, 273.

131. Searcy, R.L., (1961), In Diagnostic Biochemistry,
      p. 212, McGraw - Hill Co., New York.

132. Fletcher, A.P., Biederman, O., Moore, D., Alkjnersig,
      N., and Sherry, S., (1964) J. Clin. Invest., 43,
      681.

133. Schiff, L., M.D., (1975), In Diseases of The Liver
      p. 833 - 879.

134. Schiff, L., Rich, M.L., and Simon, S.D., (1938), Am. J. M. Si., 196, 313.

135. Hoffman, W.S., (1970), In the Biochemistry of Clinical Medicine, p. 472 Year Book, Chicago.

136. Cantarow - trumper, (1962), In Clinical Biochemistry.

137. Hall, C.A., Frame, B., and Drill, V.A., (1949), Endocrinology, 44, 76.

138. Fiske, C.H., and Subbarow, Y., (1925), J. Biol. Chem., 66 - 375.

139. Eisenthal, B., Cornish - Bowden, A., (1974), Bioch. J., 139, 715.

140. Cornish - Bowden, A. and Eisenthal, B., (1974), Bioch. J., 139, 721.

141. Jensen, R.A., and Nester, E.W., (1950), J. Biol. Chem., 186 - 557.

142. Meilands, J.B., Paul, K. Stumpf, (1964), In Out-
     lines of Enzyme Chemistry, p. 123 -127.

143. Yoshida, N.A., (1970), Clin. Chem. Acta. 30,
     546.

144. Cathelineau, I., J.M. Sandubray and C. Polonski,
     (1972), Clin. Chem. Acta. 41, 305.

145. Gerhardt, W., M. Lykkegaard Nielson, and et al.,
     (1974), Clin. Chem. Acta, 53 - 291.

146. Rassam, B.M., (1976), M.Sc., A Thesis in the
     Kinetics of LDH and other Enzymes Studies in Serum
     of Leukemic and Normal Iraqi Individuals.

147. Hardonk, M.J. and J. Koudstaal, (1968), Histo-
     chemie, 12, 18.

148. Hobbs, J.R., D.M. Campbell, and P.J. Scheuer,
     Proc. 6th. Intern. Congr. Clin. Chem. Munich.,
     1966; vol. 2, Karger, Basal / New York, (1968), p. 106.

149. Shaw, L.M., and Gray, J. (1974), Clin. Chem., 20, 4.

150. Warburg, O., and Christian, W., (1942), Biochem.
     Z., 310, 384.

151.   Bergmeyer, H.U., (1974), In Methods of Enzymatic
       Analysis, vol. II, p. 871 Academic Press, New York.

152.   Hunter, A., and Downs, C.E., (1945), J. Biol.
       Chem. 157, 427.

153.   Schwartz, M.K., O. Bodansky, (1964), Am. J. Clin.
       Path., 42, 572.

www.ingramcontent.com/pod-product-compliance
Lightning Source LLC
Chambersburg PA
CBHW080803180526
45168CB00006B/2317